Hard Fighting

Israel in Lebanon and Gaza

[handwritten inscription: 18 January 2012 — For Jim Warner — With my deepest respect and appreciation for all you do for our Nation, our Army, and our Soldiers. With all my duty,]

David E. Johnson

Prepared for the United States Army and the United States Air Force

Approved for public release; distribution unlimited

ARROYO CENTER
and PROJECT AIR FORCE

The research described in this report was sponsored by the United States Army under Contract No. W74V8H-06-C-0001 and by the United States Air Force under Contract No. FA7014-06-C-0001.

Library of Congress Cataloging-in-Publication Data

Johnson, David E. (David Eugene), 1950-
  Hard fighting : Israel in Lebanon and Gaza / David E. Johnson.
    pages cm
  Includes bibliographical references.
  ISBN 978-0-8330-5850-8 (pbk. : alk. paper)
  1. Israel. Tseva haganah le-Yisra'el. 2. Asymmetric warfare—Case studies. 3. Lebanon War, 2006. 4. Gaza War, 2008-2009. 5. United States—Military policy. I. Title.

  UA853.I8J645 2011
  956.9204'524095694—dc23

                                                                          2011049689

The RAND Corporation is a nonprofit institution that helps improve policy and decisionmaking through research and analysis. RAND's publications do not necessarily reflect the opinions of its research clients and sponsors.

Published 2011 by the RAND Corporation
1776 Main Street, P.O. Box 2138, Santa Monica, CA 90407-2138
1200 South Hayes Street, Arlington, VA 22202-5050
4570 Fifth Avenue, Suite 600, Pittsburgh, PA 15213-2665
RAND URL: http://www.rand.org/
To order RAND documents or to obtain additional information, contact
Distribution Services: Telephone: (310) 451-7002;
Fax: (310) 451-6915; Email: order@rand.org

Preface

This monograph examines the Israeli experiences in the Second Lebanon War (2006) and Operation Cast Lead (2008–2009) to assess the challenges posed by hybrid adversaries and the military capabilities needed to prevail over them. It describes what happened during the war in Lebanon, why the Israel Defense Forces (IDF) had difficulty fighting Hezbollah, what corrective measures the IDF took after Lebanon, and what happened during the operation in Gaza.

The research presented here shows that the Israeli experience provides compelling insights that will be important to the U.S. joint force—and particularly to the Air Force and the Army—as it considers the capabilities needed in the future to prevail against hybrid challenges. To this end, the monograph makes several recommendations about ways to improve the integration of air power, ground power, and intelligence, surveillance, and reconnaissance (ISR) across the Air Force and the Army in hybrid warfare.

The research reported here was sponsored by then–Major General William J. Rew, Director of Operational Planning, Policy and Strategy, Deputy Chief of Staff for Operations, Plans and Requirements, Headquarters U.S. Air Force (AF/A5X), and by Major General Francis G. Mahon, Army Quadrennial Defense Review Office, Headquarters U.S. Army (G-8). It was conducted within the Strategy and Doctrine Program of RAND Project AIR FORCE and the Strategy, Doctrine, and Resources Program of RAND Arroyo Center.

The monograph should be of interest to a wide group of U.S. Air Force, U.S. Army, and defense personnel involved in many aspects of the doctrine; combat organizations; tactics, techniques, and procedures; and related employment concepts that link interservice air and ground combat operations. It should also be of use to military and civilian planners within the Department of Defense who are working to provide the U.S. military with the capabilities needed to operate successfully across the full range of military operations, particularly against hybrid adversaries.

Questions and comments regarding this report can be addressed to the author:

David E. Johnson
RAND Corporation
1200 South Hayes Street
Arlington, Virginia 22202-5050
Phone: 703-413-1100, ext. 5205
Email: David_E_Johnson@rand.org

RAND Project AIR FORCE

RAND Project AIR FORCE (PAF), a division of the RAND Corporation, is the U.S. Air Force's federally funded research and development center for studies and analyses. PAF provides the Air Force with independent analyses of policy alternatives affecting the development, employment, combat readiness, and support of current and future aerospace forces. Research is conducted in four programs: Force Modernization and Employment; Manpower, Personnel, and Training; Resource Management; and Strategy and Doctrine.

Additional information about PAF is available on our website: http://www.rand.org/paf/.

RAND Arroyo Center

RAND Arroyo Center, part of the RAND Corporation, is a federally funded research and development center sponsored by the United States Army. Research is conducted in four programs: Strategy, Doctrine, and Resources; Force Development and Technology; Military Logistics; and Manpower and Training. The Project Unique Identification Code (PUIC) for the project that produced this document is ASPMO09224.

For more information on RAND Arroyo Center, contact the Director of Operations (telephone 310-393-0411, extension 6419; FAX 310-451-6952; email (Marcy_Agmon@rand.org), or visit Arroyo's website at http://www.rand.org/ard/.

Contents

Figures

Tables

Summary

Before the wars in Afghanistan (2001–present) and Iraq (2003–present) began, the U.S. military was largely designed for major combat operations against state actors. Over the past several years, it has made significant changes in its training, organizing, and equipping paradigms to adapt to nonstate, irregular threats in those two theaters. However, the adversaries the U.S. military has faced in these two wars have yet to employ effective standoff weaponry, such as man-portable air-defense systems (MANPADS) and antitank guided missiles (ATGMs), in large enough numbers to change how U.S. forces operate. The introduction of such weapons, if it were to occur, could radically intensify the challenges confronting U.S. forces. All that the Taliban needs to become a more lethal adversary is a state that will supply it with effective standoff weapons and training in their use.

With its current almost exclusive (and understandable) focus on irregular warfare, the U.S. joint force, particularly the U.S. Army, might be approaching a situation similar to that of the Israelis in the Second Lebanon War (2006), when the Israel Defense Forces (IDF) found itself in an unexpected "hybrid war" (defined by Frank Hoffman as a "blend of the lethality of state conflict with the fanatical and protracted fervor of irregular war"[1]). To better understand the breadth of enemy capabilities that U.S. military forces should be prepared to encounter, the U.S. Air Force and the U.S. Army asked the RAND Corporation to examine the recent experiences of the IDF in Lebanon

---

[1] Frank G. Hoffman, *Conflict in the 21st Century: The Rise of Hybrid Wars*, Arlington, Va.: Potomac Institute for Policy Studies, 2007, p. 28.

and Gaza to determine what lessons the U.S. military should learn from those conflicts.

Research included a literature review and interviews with Israeli and American military officers; Israeli and American civilian employees in the national security sector; Israeli and American academics and defense analysts; American military attachés at the U.S. Embassy in Tel Aviv; individuals in Lebanon; and Israeli, American, and British journalists. The author also assessed translations of secondary sources published in Hebrew and Arabic. The final report of the Winograd Commission (named after Eliyahu Winograd, the head of the Commission for the Examination of the Events of the 2006 Campaign in Lebanon) was of particular importance. Finally, note that some portions of this monograph draw heavily on an earlier paper on the topic published by the author in 2010.[2]

Israel's Strategic Assessment: The End of Major War?

Prior to the Second Lebanon War, three relatively recent events had influenced Israeli expectations about the future of warfare:

- The war in Kosovo (1999) and initial U.S. operations in Afghanistan (2001–present) and Iraq (2003–present) revealed the implications of the revolution in military affairs, particularly in the areas of intelligence, surveillance, and reconnaissance (ISR) and precision strike. These implications fostered a belief among some in the Israeli defense establishment that standoff attack (principally by air power) was an effective means of affecting the will of the adversary and determining conflict outcomes. Standoff attack also seemed to promise lower IDF casualties (a major domestic political consideration), less collateral damage (a key consideration for managing international and regional opinion), and cost savings.

---

[2] That work is David E. Johnson, *Military Capabilities for Hybrid War: Insights from the Israel Defense Forces in Lebanon and Gaza*, Santa Monica, Calif.: RAND Corporation, OP-285-A, 2010.

- The beginning of the Second al-Aqsa Intifada (2000–2006) forced the Israeli Army to focus on operations designed to stop terrorist attacks inside Israel, and a mindset of low-intensity conflict (LIC) developed. Furthermore, significant budget issues curtailed IDF training.
- The ongoing U.S. presence in Iraq after the end of major combat operations there, coupled with the low level of threat to Israel posed by its neighbors (except Syria), encouraged Israelis to believe that, for them, the era of major war had passed and that the role of ground forces was to carry out low-intensity irregular war.

The mindsets of Israel's military and political leaders were fundamentally—and understandably—shaped by these views. Defense spending was cut, armored-unit training (deemed largely irrelevant in LIC) was neglected, the IDF staffs and processes responsible for integrating air and ground operations were removed from brigades, and there was little training in air-ground integration. The IDF, and particularly the active Israeli Army, focused on stopping terrorist attacks; indeed, the IDF was very successful in suppressing the Second al-Aqsa Intifada and dramatically reducing Israeli casualties. However, the Israeli Army's almost exclusive focus on LIC resulted in a military that was generally incapable of executing the integrated joint air-ground-ISR operations associated with major combat.[3] Unfortunately for the Israelis, the Second Lebanon War (2006) demonstrated that taking a defended position from a hybrid force that is armed with standoff fires (e.g., ATGMs, mortars, MANPADS, rockets) requires forces trained and organized for integrated fire and maneuver tactics, however reduced in scale (compared with major combat operations) the size of that hybrid force might be.

---

[3] This monograph broadly defines *air-ground-ISR capabilities* as the constellation of air, ground, and space means used to find, fix, and capture or kill an adversary. To be truly effective, these capabilities must be integrated across services (and agencies), and the relevant information provided by the wide array of ISR resources must be readily available at the levels that need it.

The Second Lebanon War: A Wake-Up Call

Since its inception in the early 1980s, Hezbollah has conducted raids and suicide attacks against IDF troops and against targets in Lebanon to eliminate "colonialist" influences in Lebanon and to establish an Islamic regime. On July 12, 2006, Hezbollah operatives ambushed two IDF High-Mobility Multi-Purpose Wheeled Vehicles (HMMWVs) conducting a routine patrol along the border between Israel and Lebanon, taking two soldiers hostage. This action led to the first Israeli military operation in Lebanon since the IDF's withdrawal in 2000.

Southern Lebanon offers decided advantages to the defender: Its complex terrain frequently forces military vehicles to remain on roads, thus increasing opportunities for ambush and for attacks with improvised explosive devices (IEDs) and mines. Hezbollah itself also proved an unexpectedly formidable adversary. During the years leading up to the Second Lebanon War, Hezbollah forces received extensive training in Lebanon, Syria, and Iran and learned how to blend guerilla tactics with conventional military tactics and weapons to create an innovative concept for defending southern Lebanon from an Israeli incursion. Hezbollah organized military units to conduct decentralized operations, built well-equipped bunkers across southern Lebanon, stockpiled supplies, and armed itself with effective standoff weapons (including ATGMs, rocket-propelled grenades, MANPADS, mortars, and a wide array of rockets).

The IDF initially expected to achieve its objectives—getting Lebanon to control Hezbollah and securing the return of the abducted Israeli soldiers—largely through air and artillery strikes and limited ground raids. Israel's political and military leaders were against the idea of deploying a large ground force and calling up reserves, but they ended up doing just that when IDF standoff fires did not result in success. However, the IDF had no accepted operational plan, and the ground war was improvised. Conditioned for LIC, IDF ground forces encountered real difficulties when they confronted Hezbollah, and they paid a heavy price in casualties for their lack of preparation to confront a hybrid opponent. After 34 days, a ceasefire agreement went into effect, but the war was highly problematic for Israel. This assessment is reflected in the report of the Winograd Commission tasked with examining the war.

In spite of stubborn, brave fighting by many IDF troops, regular and reserve forces alike, the Israeli military as a whole, and the Israeli Army in particular (in contrast with the Israeli Air Force and, in part, with the Israeli Navy), failed to fulfill many of its missions. Despite its significant quantitative and qualitative edge, the IDF not only failed to vanquish Hezbollah but also failed to halt Hezbollah-launched rocket fire into Israel, which continued until the last day of the war.[4] Furthermore, approximately 120 IDF forces were killed (and more than 1,000 were wounded) in the war, in addition to the 37 Israeli civilians who died (mostly in rocket attacks).

Hezbollah and Lebanese civilian losses were also high: Estimates suggest that 250–800 Hezbollah forces and 900–1,100 Lebanese civilians were killed. Nevertheless, just by surviving and showing that it could continuously launch rockets at Israel, Hezbollah was able to claim victory. The IDF's reputation as an invincible military—a reputation fundamental to its ability to deter potential adversaries—was severely tarnished.

Operation Cast Lead: Back to Basics

In the aftermath of the Second Lebanon War, the IDF went "back to basics," drawing up a new defense plan that emphasized building up ground forces and training them extensively in high-intensity conflict (HIC) skills, particularly combined-arms fire-and-maneuver tactics. Before the Second Lebanon War, roughly 75 percent of IDF training concerned LIC, and just 25 percent concerned HIC. After the war, the IDF decided to devote 80 percent of training to high-intensity combined-arms training. In the regular forces, training time was doubled, and combined-arms, live-fire exercises were instituted for brigade combat teams. The IDF began adding new heavy infantry armored

---

[4] The Commission for the Examination of the Events of the 2006 Campaign in Lebanon, *The Second Lebanon War: Final Report*, Vol. I, January 2008, pp. 232–233.

fighting vehicles (Namers) and additional tanks (Merkava IVs), and it resumed armored-force and reserve training. The Israeli Army and the Israeli Air Force also markedly increased mutual cooperation in the following realms: ISR, the integration of unmanned aerial vehicles (UAVs), and close air support. The Israeli Air Force also returned tactical air-control capabilities to Israeli Army brigades and worked closely with the Israeli Army to improve air-ground operations.

When the IDF went into the Gaza Strip in December 2008, it was markedly better prepared to fight another hybrid opponent: Hamas. Hamas had been engaged in an aggressive military buildup since the Israeli withdrawal from Gaza in 2005, increasing its force to more than 10,000 operatives, improving its command and control (C2) capabilities, developing an extensive network of tunnels to use as bunkers and traps, and procuring standoff weaponry (including short- and intermediate- range rockets, ATGMs, and a small number of MANPADS).

Hamas increased the number of rocket and mortar attacks it launched from Gaza into Israel, particularly after it gained control in Gaza. On November 4, 2008, the IDF launched Operation Double Challenge against Hamas tunnels. After that, Hamas stepped up its attacks, launching more than 700 rockets and mortars, including several longer-range Grads, into Israel.[5] In response, Israel planned a campaign that included all elements of its conventional military power: air, ground, navy, intelligence, and the reserves. On December 27, 2008, the IDF began Operation Cast Lead with a massive air attack on Hamas. Unlike the Second Lebanon War, Operation Cast Lead included, from the beginning, plans for a ground campaign and for reserve mobilization. IDF artillery and air strikes paved the way for ground maneuver by brigade combat teams, hitting Hamas positions and detonating mines and IEDs. IDF engineers used armored D-9 bulldozers to clear paths through the remaining IEDs. Armored units composed of tanks and armored personnel carriers were also a key component of the operations, providing protected, mobile precision firepower and intimidating the enemy.

---

[5]   Israel Defense Forces, "Rocket Statistics, 3 Jan 2009," web page, January 3, 2009. This web page also notes that 7,200 rockets and mortars were launched from Gaza at Israel between 2005 and 2008.

The IDF firepower that preceded the ground attack, and the rapidity with which the maneuver was carried out, surprised Hamas, and it was driven out of its generally well-organized and prepared positions and back to improvised positions. Although the IDF did not put a complete halt to rocket launches out of Gaza into Israel, it was able to decrease their number. Even considering that the terrain in Gaza was more conducive to IDF operations than the terrain in southern Lebanon, and that Hamas was not as formidable an opponent as Hezbollah, Israel successfully demonstrated its renewed competence in air-ground operations—a demonstration key to restoring its military deterrent.

The "Middle" of the Range of Military Operations

The Second Lebanon War showed that the IDF had a significant capability gap in "the middle." As shown in Figure S.1, there are opponents with three basic levels of military competence. Each level places different demands on the military forces designed to confront those adversaries.

Importantly, competent nonstate actors can easily transition from the low end (nonstate irregular) to the middle (nonstate hybrid). All that is needed is a state sponsor that provides weapons and training to the irregular forces. The United States itself enabled such a transition in Afghanistan in the 1980s when it gave effective standoff capabilities (including Stinger missiles) and training to the mujahedeen, helping them upgrade from an irregular force into a hybrid adversary—an upgrade that forced changes in Soviet tactics, operations, and strategy in Afghanistan. Thus, an important strategic consideration is how to deter state actors from providing capabilities to irregular actors, thus preventing the latter from becoming hybrid actors.

The military capabilities needed to deal with hybrid opponents are similar to those the IDF deployed during Operation Cast Lead and during initial air strikes in the Second Lebanon War. Table S.1 summarizes the roles of air power and ground power in dealing with the three broad levels of conflict.

Figure S.1
Levels of Adversaries and Their Associated Military Capabilities, with Examples

- Mujahedeen (Afghanistan, 1979)
- PLO (West Bank/Gaza, 2001)
- Al-Qaeda in Iraq (2007)
- Taliban (Afghanistan, 2009)

Nonstate Irregular

- *Organization:* not well trained; little formal discipline; cellular structure; small formations (squads)
- *Weapons:* small arms; RPGs; mortars; short-range rockets; IEDs/mines
- *Command and control:* cell phones; runners; decentralized

- Mujahedeen (Afghanistan, 1988)
- Chechen militants (Chechnya, 1990)
- Hezbollah (Lebanon, 2006)
- Hamas (Gaza, 2008)

State-Sponsored Hybrid

- *Organization:* moderately trained; disciplined; moderate-sized formations (up to battalion)
- *Weapons:* same as irregular, but with standoff capabilities (ATGMs, MANPADS, longer-range rockets)
- *Command and control:* multiple means; semicentralized

- Soviet Union (Afghanistan, 1970s–1980s)
- Russia (Chechnya, 1990s)
- Israel (Lebanon, 2006)
- Georgia (Georgia, 2008)
- Russia (Georgia, 2008)
- Israel (Gaza, 2008)
- United States (Afghanistan, Iraq, 2010)

State

- *Organization:* hierarchical; brigade- or larger-sized formations
- *Weapons:* sophisticated air defenses; ballistic missiles; conventional ground forces; special operations forces; air forces; navies; some have nuclear weapons
- *Command and control:* all means; generally centralized

SOURCES: Johnson, *Military Capabilities for Hybrid War*; David E. Johnson and John Gordon IV, Observations on Recent Trends in Armored Forces, Santa Monica, Calif.: RAND Corporation, OP-287-A, 2010.
NOTE: PLO = Palestine Liberation Organization. RPG = rocket-propelled grenade.
RAND MG1085-S.1

Table S.1
Air and Ground Power Across the Levels of Adversary Capabilities

	Irregular Warfare	State-Sponsored Hybrid Warfare	Deterrence/Major Combat Operations
The role of air power	• Overhead ISR and signals intelligence are crucial because the enemy does not mass. These capabilities are critical to finding and attacking high-value targets. • Air mobility is critical to supply and evacuation. • Air power is used mostly in ground-directed close air support with tight rules of engagement. It is key for force protection in extremis. • Air superiority is contested below 3,000 feet. Above 3,000 feet, air power is mainly invulnerable.	• Overhead ISR and signals intelligence are tightly linked with precision strike. • Air power is critical to attacking the enemy's deep strike assets and high-value targets. • Tight rules of engagement for centralized strikes and close air support are required. • Air power is used for the suppression of enemy standoff systems to support (complement) ground maneuver. • Air power complicates the enemy's ability to mass and be reinforced. • Air superiority may be contested below 20,000 feet.	• Air power is critical to deterrence achieved through global reach and strike capabilities. • Overhead ISR and signals intelligence are tightly linked with precision strike. • Strategic and operational air mobility and tankers are critical capabilities. • Air and space superiority may be contested at all levels. • Centralized control is critical. • Air power precludes large-scale ground maneuver by the enemy. • Air bases may be contested. • Operations may occur in a nuclear, biological, or chemical environment.

Table S.1—Continued

	Irregular Warfare	State-Sponsored Hybrid Warfare	Deterrence/Major Combat Operations
The role of ground power	• Ground power is focused on establishing security, obtaining human intelligence, and training indigenous forces. • Maneuver is focused on clearing, holding, and building. • Dispersed operations increase the difficulty of force protection. • Tight rules of engagement demand rigorous target identification.	• Ground power is critical to forcing enemy reaction and to uncovering hidden assets. • Combined-arms operations are fundamental to success. • Ground power closes with enemy forces. • Ground power conducts decentralized operations against dispersed adversaries. • High-intensity MCO-like operations are possible at the brigade level and below. • Lines of communication may be vulnerable.	• Troop deployment is a key signal of national commitment. • Combined-arms operations are the key to success. • Ground maneuver forces an operational reaction from the enemy. • Ground power engages ground units that avoid air attacks and indirect fire. • Ground power is critical for exploiting operational opportunities and pursuing enemy forces. • Ground power deals with hybrid or irregular threats. • Ground power is critical to establishing post-MCO security and stability. • Basing and staging may be contested. • Operations may occur in a nuclear, biological, or chemical environment.
The level of air-ground integration	• Operations are ground-centric but highly dependent on air power. • C2ISR and joint tactical air controllers are best integrated at lower echelons for direct support.	• Balanced operation requires tighter coordination and extensive training and rehearsals. • Integration ensures high responsiveness to ground units and integration at levels below the theater.	• Air power control is highly centralized. • Air superiority is critical to ground maneuver. • Integrated suppression of enemy air defenses is key. • Supported-supporting relationships depend on the operation; the air or ground commander could lead. Leadership could change during an operation.

NOTE: C2ISR = command and control, intelligence, surveillance, and reconnaissance. MCO = major combat operations.

Lessons and Recommendations

The ultimate lesson from the Israeli experiences in Lebanon and Gaza is this: An enemy's capabilities will largely determine the war a nation will have to fight. The imperative to conduct protracted low-intensity operations in Iraq and Afghanistan with limited ground forces has required the United States to make choices about training focus and combat preparation that have diminished the readiness of U.S. military forces to prevail against a hybrid enemy. In particular, the skills and processes required for integrated air-ground-ISR operations need to be assessed and strengthened. To that end, this monograph highlights the following key implications from the Israeli experiences in Lebanon and Gaza:

- Despite the smaller scale of the conflict, such hybrid opponents as Hezbollah and Hamas constitute a challenge that is qualitatively similar to that posed by major combat operations because of their training, discipline, organization, command and control, and effective standoff weapons (e.g., ATGMs, MANPADS, surface-to-surface rockets). These capabilities are "game changers": Irregular opponents who attain them can rapidly ratchet up the intensity level of a conflict, and defeating these opponents requires different skills than those used in counterinsurgency operations. After years of focusing on LIC operations in Gaza and the West Bank, the IDF (particularly the Israeli Army) was not prepared for the challenges posed by Hezbollah. The U.S. military faces similar issues after years of focusing on irregular warfare in Afghanistan and Iraq.
- There are no single-service solutions to the challenges posed by hybrid adversaries. Israel's training, organizational, and doctrinal changes after the Second Lebanon War, and particularly changes in air-ground-ISR integration, paid off in Operation Cast Lead for the IDF. Similar changes across the doctrine, organization, training, materiel, leadership and education, personnel, and facilities (DOTMLPF) spectrum may be necessary to prepare the U.S. joint force for hybrid opponents.

- Precision standoff fires were critical to—but not sufficient for—coping with hybrid opponents in Lebanon and Gaza, particularly when Israel's adversaries were operating among the population. Thus, as the IDF realized, joint, combined-arms operations are crucial in finding, fixing, and capturing or killing hybrid adversaries who are dispersed and concealed in complex terrain. Furthermore, because of the fleeting nature of the enemy, ground force brigades, rather than higher echelons, were the locus of decentralized tactical decisionmaking and combined-arms action in Gaza.

- Hybrid opponents (e.g., Hamas) become more visible when they take over and move into government buildings; this makes them more susceptible to precision strike. Similarly, intermediate- and long-range rockets are easier to find and destroy because of their size and the requirement that they be launched from relatively open sites. Thus, the Israeli Air Force (IAF) was very successful in finding and destroying intermediate- and long-range rockets and in attacking infrastructure targets. This is a unique capability afforded by air power, and it was particularly important in preempting the use of more-capable rockets. Additionally, only fixed-wing aircraft were capable of delivering precision ordnance with the payloads necessary to destroy large, multistory structures and tunnels.

- Persistent ISR coverage is crucial in tracking mobile opponents (particularly mortar, rocket, and ATGM crews) and high-value targets (e.g., intermediate- and long-range rockets in Lebanon, key enemy leaders). UAVs are particularly valuable because of their loitering time and because using unmanned (rather than manned) aircraft in high-threat environments eliminates the risk of losing a pilot. The ability to command and control a combination of ground forces, fixed-wing aircraft, attack helicopters, UAVs, and other assets—all operating off a "common picture" of the battlefield—is critical in attacking fleeting, time-sensitive targets and in avoiding collateral damage.

- Highly dispersed, low-signature targets (e.g., short-range rockets, ATGMs) are difficult to find and attack by air-only means, and attacking them requires comprehensive air-ground-ISR integration at low (i.e., battalion and brigade) levels; common references (e.g., detailed maps and control measures); and a shared, real-time ISR picture.

- Successfully striking targets "amongst the people" in Gaza required a combination of exquisite interagency intelligence, precision strike, and low-yield weapons.[6] These capabilities were needed to identify targets, warn civilians, attack hidden targets (e.g., weapons caches, rockets) and avoid killing the wrong people. This level of intelligence may not be attainable by the Israelis elsewhere (e.g., in Lebanon) or by the United States in current and future conflict environments.

- Armored forces based on tanks and armored personnel carriers are key elements of any force that will fight hybrid enemies with a modicum of training, organization, effective standoff weapons (e.g., ATGMs, MANPADS), IEDs, and mines. Light and medium forces (e.g., Stryker brigade combat teams in the U.S. Army) can complement armored forces, particularly in urban and other complex terrain, but they do not provide the survivability, lethality, or mobility inherent in armored forces. Quite simply, armored forces reduce operational risks and minimize friendly casualties. Information cannot replace armor.

In light of these insights, this monograph presents the following lessons and recommendations:

- The skills and processes needed to prevail against hybrid opponents are different in many respects from those required to conduct a counterinsurgency. They require the highly integrated joint, combined-arms fire-and-maneuver skills used in major combat operations, but at a lower organizational level (i.e., the brigade combat team level). The skills and processes needed for integrated air-ground-ISR operations against hybrid adversar-

---

[6] The phrase *amongst the people* is from a chapter titled "War Amongst the People" in Rupert Smith, *The Utility of Force: The Art of War in the Modern World*, New York: Vintage Books, 2005.

ies with effective standoff fires capabilities—skills and processes that may have atrophied among U.S. forces during the United States' protracted counterinsurgency operations in Afghanistan and Iraq—need to be assessed and strengthened in both the U.S. Air Force and the U.S. Army. Therefore, the combat training centers should continue their renewed emphasis on preparing forces for the full range of military operations and should incorporate hybrid operations and hybrid opposing forces into training exercises and experiments.

- U.S. Air Force and U.S. Army training, organization, and equipping efforts need to prepare forces for hybrid challenges, which could materialize in Afghanistan or elsewhere with the introduction of effective standoff weapons, particularly MANPADS. The appearance of these types of weapons in any numbers in Afghanistan could radically change how U.S. forces operate.

- The inherent nature of hybrid threats requires detailed air-ground-ISR integration, and the U.S. Air Force can make an important contribution by ensuring that it has a highly capable and senior presence in brigade staffs and in subordinate maneuver forces and staffs. In the IDF, C2 during the air-ground phase of Operation Cast Lead was pushed down to the brigade level. Intelligence, fires, and maneuver were fused because of the fleeting nature of targets, the complex operating environment, the dispersed nature of the adversary, and the need to limit collateral damage. It is reasonable to assume that the future hybrid threat environments that the U.S. joint force might face will have similar characteristics. Furthermore, the U.S. Army has adopted a brigade combat team structure not unlike that used by the IDF in Operation Cast Lead. Therefore, the U.S. Air Force and the U.S. Army should assess what kind of air operations planning and C2 capabilities should reside in U.S. Army brigades and consider the integration of capabilities from across the other military services (e.g., the U.S. Navy and U.S. Marine Corps) and federal agencies (e.g., the Central Intelligence Agency, the National Security Agency).

- The threats posed by precision standoff fire systems (e.g., MANPADS, ATGMs) are different than those posed by irregular opponents. Technological countermeasures (e.g., jammers, active armor)—as well as tactics, techniques, and procedures—are needed to defeat these weapons.

- The U.S. joint force needs to develop and institutionalize processes to integrate and control cross-service platforms and capabilities at the level of the brigade combat team. This will be harder in the U.S. joint force than it is in the IDF because, in the latter military, the IAF owns and operates virtually every air platform (i.e., fixed-wing aircraft, rotary-wing aircraft, and most UAVs). However, integration will become more complex in the IDF as UAVs continue to proliferate in the ground forces.

- The U.S. joint force needs to develop common reference systems for urban terrain that enable rapid mutual target recognition, mensuration, and attack. These systems could include predetermined common coordinates and numbering systems for buildings.

- The U.S. joint force needs the capability to find, collaboratively observe, and strike fleeting targets among civilian populations with precision and very-low-yield weapons. This has been done in ad hoc ways in Iraq (in, for example, the 2008 battle in Sadr City) but has not yet been institutionalized (as it was in the IDF before Operation Cast Lead).

- The U.S. Air Force needs to have the capabilities required to destroy large structures (e.g., multistory buildings) and subterranean complexes like those used by Hezbollah and Hamas, but it must also be able to limit civilian casualties and collateral damage. In all likelihood, challenges in this area will increase as adversaries dig deeper and continue to operate in urban areas. Additionally, the realities of fighting in complex terrain, particularly urban areas, can require forces to drop munitions closer to friendly troops. Thus, smaller and highly precise munitions are needed to avoid fratricide.

- To effectively engage hybrid opponents, an air-ground-ISR team must receive detailed training and conduct rehearsals. Therefore, to avoid ad hoc arrangements that limit effectiveness and replicability, the U.S. Air Force and the U.S. Army should examine existing arrangements for joint planning and execution and consider the possibility of establishing habitual relationships between air and ground forces within a theater of operations. As the IDF experience in Gaza shows, trust between air and ground forces is a combat multiplier, and only through habitual association and personal relationships can this trust be truly established.

Israel's experiences in Lebanon and Gaza show that hybrid opponents can create significant challenges for nations whose ground forces are focused on irregular warfare and whose air forces are designed to maintain a high-end deterrent and warfighting capability. The Israelis learned the hard way in Lebanon that there was a gap in the IDF's ability to carry out operations "in the middle." As the U.S. joint force prepares to confront the full spectrum of potential future challenges, Israel's experiences are well worth learning from.

Acknowledgments

Many people contributed their time and intellectual energy to the evolution of this monograph. Two of my RAND colleagues—Mike Spirtas and Ghassan Schbley—made important contributions to the draft manuscript, and their contributions are reflected in this volume.

Additionally, I would like to acknowledge the many individuals in the Israel Defense Forces, the Israel Ministry of Foreign Affairs, the Embassy of the United States in Israel, the Embassy of Israel in the United States, the U.S. Department of Defense, and the U.S. Joint Chiefs of Staff who gave freely of their time during interviews and discussions.

I owe a particular debt of gratitude to Brigadier General Itai Brun and his staff at the IDF Dado Center for Interdisciplinary Studies, Lieutenant Colonel Ron Amir (Israel Air Force), Eitan Shamir (Dado Center), Colonel Meir Finkel (Israel Ground Forces), Colonel Avi Gil (Israel Ground Forces), Major Uzi Ben-Shalom (Israel Ground Forces), Brigadier General (Reserves) Ephraim Segoli (The Fisher Brothers Institute for Air and Space Strategic Studies), and Brigadier General (Reserves) Asaf Agmon (The Fisher Brothers Institute for Air and Space Strategic Studies). These individuals sponsored my visits to Israel and helped arrange interviews and meetings. Dima Adamsky, Colonel Amir, Colonel Finkel, and William F. Owen read the draft manuscript and provided very useful comments.

I also appreciate the interest and support provided by then–Lieutenant Colonel Vincent Alcazar (Headquarters, U.S. Air Force, A5XS) and Timothy S. Muchmore (Headquarters, U.S. Army, G-8).

Finally, each of the monograph's three reviewers—Colonel Gian Gentile, Adam Grissom, and Karl Mueller—provided thorough and very useful reviews, and Erin-Elizabeth Johnson did a masterful editing job.

The efforts of all these individuals contributed immeasurably to the final product that is this monograph. I owe them all an enormous debt of gratitude.

Abbreviations

ATGM	antitank guided missile
C2	command and control
C2ISR	command and control, intelligence, surveillance, and reconnaissance
COIN	counterinsurgency
COMINT	communications intelligence
COS	IDF Chief of Staff
DoD	Department of Defense
DOTMLPF	doctrine, organization, training, materiel, leadership and education, personnel, and facilities
EBO	effects-based operations
ERA	explosive reactive armor
FCS	U.S. Army Future Combat Systems
GPS	Global Positioning System
HIC	high-intensity conflict
HMMWV	High Mobility Multi-Purpose Wheeled Vehicle
HUMINT	human intelligence
HVT	high-value target

IAF	Israeli Air Force
IDF	Israel Defense Forces
IED	improvised explosive device
IR	Islamic Resistance
ISR	intelligence, surveillance, and reconnaissance
IW	irregular warfare
JFAC	joint forward air control
JTAC	joint tactical air control
LIC	low-intensity conflict
MANPADS	man-portable air-defense system
MBT	main battle tank
MCO	major combat operations
MRAP	mine-resistant ambush-protected vehicle
NIS	new Israeli shekel
OTRI	Operational Theory Research Institute
PA	Palestinian Authority
PGM	precision guided munition
PIJ	Palestinian Islamic Jihad
PLO	Palestine Liberation Organization
RMA	revolution in military affairs
RPG	rocket-propelled grenade
SEAD	suppression of enemy air defenses
SIGINT	signals intelligence
SOD	Systemic Operational Design

SSM	surface-to-surface missile
TTP	tactics, techniques, and procedures
UAV	unmanned aerial vehicle
UN	United Nations

CHAPTER ONE

Introduction

Since 2003, the United States has been deeply engaged in counterinsurgency (COIN) campaigns in Afghanistan and Iraq, and the U.S. military has made significant changes to its training, organizing, and equipping paradigms—once heavily biased toward major combat operations—to adapt to these two irregular wars. Indeed, a 2008 U.S. Department of Defense (DoD) directive stated that it is DoD policy to "recognize that IW [irregular warfare] is as strategically important as traditional warfare."[1] This recognition followed the revamping of U.S. strategy in Iraq, publication of a new U.S. Army and U.S. Marine Corps COIN manual, publication of a U.S. Air Force IW manual, and, most recently, a reexamination of U.S. strategy in Afghanistan.[2]

The conditions within Iraq and Afghanistan that shaped the COIN strategies in each country were very different, but they shared one essential characteristic: The adversaries in both of these wars never employed effective standoff weaponry—e.g., man-portable air-defense systems (MANPADS), antitank guided missiles (ATGMs), anti-ship missiles, and surface-to-surface intermediate- or long-range rockets— in any significant numbers. Fortunately, coalition forces have thus far

---

[1] U.S. Department of Defense, *Irregular Warfare*, DoDD 3000.07, December 1, 2008.

[2] See Jesse Lee, "A New Strategy for Afghanistan and Pakistan," post on the White House Blog, March 27, 2009; Stanley A. McChrystal, "COMISAF's Initial Assessment," Headquarters, International Security Assistance Force, Kabul, Afghanistan, August 30, 2009; U.S. Department of the Air Force, *Irregular Warfare*, AFDD 2-3, Washington, D.C., 2007; and U.S. Department of the Army and Marine Corps Combat Development Command, U.S. Department of the Navy, *Counterinsurgency*, FM 3-24/MCWP 3-33.5, 2006.

faced irregular opponents equipped only with small arms, machine guns, rocket-propelled grenades (RPGs), mortars, improvised explosive devices (IEDs), and a limited number of both short-range rockets and MANPADS.[3] The nature of the enemy in Afghanistan and Iraq has allowed U.S. forces to focus almost exclusively on COIN and counter-terrorism operations.

Clearly, however, the duration, intensity, and geographic scale of the wars in Afghanistan and Iraq have placed extraordinary demands on all of the services. The U.S. Army has been under the greatest strain, with former U.S. Army Chief of Staff General George W. Casey noting,

> The Army is out of balance. The current demand for our forces in Iraq and Afghanistan exceeds the sustainable supply and limits our ability to provide ready forces for other contingencies. Even as the demand for our forces in Iraq decreases, the mission in Afghanistan and other requirements will continue to place a high demand on our Army for years to come. Current operational requirements for forces and insufficient time between deployments require a focus on counterinsurgency training and equipping to the detriment of preparedness for the full range of military missions. Overall, we are consuming readiness as fast as we can build it.[4]

The wars in Afghanistan and Iraq have also shaped U.S. Air Force programs and capabilities. Secretary of the Air Force Michael B. Donley and Chief of Staff Norton A. Schwartz testified before Congress that the U.S. Air Force is

> also responding to significant growth in the requirements for Irregular Warfare (IW) capabilities with major investments in

---

[3]   There have been reports of MANPADS and ATGM use in Iraq and Afghanistan, but their use has been very infrequent and is not representative of a military capability like that possessed by Hezbollah. Additionally, Hezbollah had not only large quantities of ATGMs but also some of high quality (e.g., the AT-14 Kornet).

[4]   Pete Geren and George W. Casey, *2009 Army Posture Statement*, May 2009.

special operations airlift, close air support and Intelligence, Surveillance and Reconnaissance (ISR).[5]

However, because of its role, shared with the U.S. Navy, as the nation's principal nuclear and long-range strike deterrent force, the U.S. Air Force faces a broader problem than the U.S. Army. Growth in the U.S. Air Force's IW capabilities is competing with its traditional programs and, as Secretary Donley remarked, is causing concerns about balance:

> Strategic balance is a key element for consideration. Our modernization and recap strategy must reflect a balance of low end vs. high end, and of today's fight vs. tomorrow's challenges. At the highest end we need to support nuclear deterrence—still a critical element of our security strategy.[6]

Thus, the U.S. Army has been focused almost exclusively on the low end of the spectrum of conflict because of the demands for ground forces in Iraq and Afghanistan, whereas the U.S. Air Force's attention is split between the low and high ends of the range of military operations.

The United States faces a situation that is similar to that Israel faced when it fought the Second Lebanon War (2006). In that war, an Israeli military that was organized, trained, and equipped to fight low- and high-intensity wars confronted Hezbollah, an adversary that fell between those extremes. Hezbollah practiced what has become known as *hybrid war*, defined by Frank Hoffman as a "blend of the lethality of state conflict with the fanatical and protracted fervor of irregular war."[7] As this monograph demonstrates, the Israelis were not prepared for the challenges posed by Hezbollah. The current U.S. focus on IW

---

[5] Michael B. Donley and Norton A. Schwartz, *United States Air Posture Statement: 2009*, May 19, 2011.

[6] Michael B. Donley, "Air Force Modernization and Recapitalization Strategy," keynote address prepared for the Aerospace Industries Association, November 18, 2008.

[7] Frank G. Hoffman, *Conflict in the 21st Century: The Rise of Hybrid Wars*, Arlington, Va.: Potomac Institute for Policy Studies, 2007, p. 14.

and COIN is having a similar deleterious effect on the U.S. joint force's ability to contend with hybrid adversaries.[8]

In the aftermath of Lebanon, the Israel Defense Forces (IDF) undertook significant reforms whose effectiveness was demonstrated in Gaza during Operation Cast Lead (December 2008–January 2009). The measures the IDF adopted to improve its ability to deal with hybrid opponents are highly relevant to the U.S. joint force.

Purpose

The purpose of this manuscript is to assess

- the state of the Israeli military before the Second Lebanon War
- the challenges that hybrid warfare posed for the IDF in southern Lebanon
- what the Israelis learned from the experience and how they adapted to fight in Gaza
- the lessons that the U.S. military can learn from Israeli experiences.

Both the Second Lebanon War and Operation Cast Lead provide insight into the challenges that nonstate actors with a modicum of training, organization, and discipline—and effective standoff weapons—can pose. These are challenges that the United States too will likely face, since the proliferation of relatively low-cost standoff weaponry makes it highly probable that hybrid threats will continue

---

[8]   General Casey has frequently remarked on the issue of having to prepare for COIN at the expense of full-spectrum capability. See U.S. Department of the Army, *2008 Army Posture Statement: A Campaign Quality Army with Joint and Expeditionary Capabilities*, Washington, D.C., 2008, p. 6, where General Casey notes that "current operational requirements for forces and insufficient time between deployments require a focus on counterinsurgency training and equipping to the detriment of preparedness for the full range of military missions." For an assessment of the effect of the focus on COIN on the U.S. Army's fires support system, see Sean MacFarland, Michael Shields, and Jeffrey Snow, "White Paper for CSA: The King and I—The Impending Crisis in Field Artillery's Ability to Provide Fire Support to Maneuver Commanders," undated [2008].

into the future and that U.S. joint forces will have to be prepared for this type of warfare.

Methodology

Research included a program of interviews spanning almost three years. Among those interviewed were Israeli and American military officers; Israeli and American civilian employees in the national security sector; Israeli and American academics and defense analysts; American military attachés at the U.S. Embassy in Tel Aviv; individuals in Lebanon; and Israeli, American, and British journalists. The author also participated in several seminars in Israel and the United States that focused on air-ground operations and hybrid warfare, and he reviewed translations of secondary sources published in English (as well as materials translated from Hebrew and Arabic into English). Of particular importance to understanding the IDF before and during the Second Lebanon War was the English translation of the final report of the Commission for the Examination of the Events of the 2006 Campaign in Lebanon, commonly known as the Winograd Commission (named after Eliyahu Winograd, the head of the commission). Finally, note that some portions of this monograph draw heavily on an earlier paper on the topic published by the author in 2010.[9]

The methodological approach focused principally on identifying military lessons from the IDF's recent experiences in Lebanon and Gaza. Although ever mindful of Carl von Clausewitz's dictum that "the political object is the goal, war is the means of reaching it, and the means can never be considered in isolation from their purpose,"[10] the examination is mainly focused on military capabilities because it is those capabilities that provide the instrument central to the "means"

---

[9] That work is David E. Johnson, *Military Capabilities for Hybrid War: Insights from the Israel Defense Forces in Lebanon and Gaza*, Santa Monica, Calif.: RAND Corporation, OP-285-A, 2010.

[10] Carl von Clausewitz, *On War*, Michael Howard and Peter Paret, eds., trans., Princeton: Princeton University Press, 1984, p. 87.

of war. Clearly, if the military instrument is faulty, war as a means to obtain political objectives becomes problematic. Thus, this monograph's discussions of political considerations and strategy are generally limited to how they shaped military decisions, capabilities, and operations.

The difficulties inherent in conducting comprehensive research about recent and ongoing Israeli operations at an unclassified level were ever present. The observations of Sergio Catignani, a scholar who researches the IDF, are quite relevant in this regard:

> There are major difficulties in studying the IDF and Israeli security in general. Yoram Peri indicated the crux of such difficulties when writing that: "The all-encompassing nature of war in Israel and the centrality of security to national existence have created a situation whereby numerous spheres . . . fall within the security ambit and are enveloped in secrecy." So ingrained is the secretive mind-set of the Israeli security establishment that native researchers with ties to the IDF have stated that even data on the Israeli reserve army is hard to access or find.[11]

The difficulties encountered in attempting to uncover specifics about Hezbollah's military strategy and operations are even more daunting, as Alastair Crooke and Mark Perry note in an article about the Second Lebanon War:

> The portrait that we give here is also limited. Hezbollah officials will neither speak publicly nor for the record on how they fought the conflict, will not detail their deployments, and will not discuss their future strategy.[12]

Despite these organic research limitations—which, to some degree, attend any project that examines a national security organization in depth, regardless of its national origin—quite a bit was learned

---

[11] Sergio Catignani, *Israeli Counter-Insurgency and the Intifadas: Dilemmas of a Conventional Army*, London: Routledge, 2008, p. 13.

[12] Alastair Crooke and Mark Perry, "How Hezbollah Defeated Israel: Part 1—Winning the Intelligence War," ATimes.com, October 12, 2006.

about the IDF. However, considerably less was learned about Hezbollah, which is an acknowledged weakness of the study.

Monograph Organization

Chapters Two and Three examine the Second Lebanon War and Operation Cast Lead, respectively, with each chapter describing the basis of the conflict, the state of the IDF, the challenge posed by the opponent, and the way the operation unfolded. Each also offers an assessment of IDF performance. In Chapter Four, insights from both conflicts that are relevant to the U.S. Air Force, the U.S. Army, and the broader U.S. joint force are identified and explained.

CHAPTER TWO

The Second Lebanon War

In summary, when the strongest military in the Middle East embarked to fight the Hezbollah and does not clearly defeat it, this had far-reaching adverse consequences for Israel's status.[1]

On July 12, 2006, Hezbollah operatives ambushed two IDF High Mobility Multi-Purpose Wheeled Vehicles (HMMWVs) conducting a routine patrol along the border between Israel and Lebanon near the village of Zarit (shown on the map in Figure 2.1).[2] Three IDF soldiers were injured and at least three others were killed. Two soldiers, Udi Goldwasser and Eldad Regev, were taken from the ambush site into Lebanon. This action led to the first Israeli military operation in Lebanon since the IDF's withdrawal in 2000 and to the largest-scale Israeli military action since the First Lebanon War (1982). This chapter discusses the factors that contributed to the conflict, the state of the Israeli military, the state of Hezbollah's military forces, the events of the war, and the effect of the conflict on the IDF.

---

[1] The Commission for the Examination of the Events of the 2006 Campaign in Lebanon, *The Second Lebanon War: Final Report*, Vol. I, January 2008, p. 76.

[2] The best account of this action, and of the Second Lebanon War in general, is found in Amos Harel and Avi Issacharoff, *34 Days: Israel, Hezbollah, and the War in Lebanon*, New York: Palgrave Macmillan, 2008. For more detail on the events of July 12, see Nicholas Blanford, "Deconstructing Hizbullah's Surprise Military Prowess," *Jane's Intelligence Review*, November 1, 2006; and Greg Myre and Steven Erlanger, "Clashes Spread to Lebanon as Hezbollah Raids Israel," *The New York Times*, July 12, 2006.

Figure 2.1
Map of Southern Lebanon

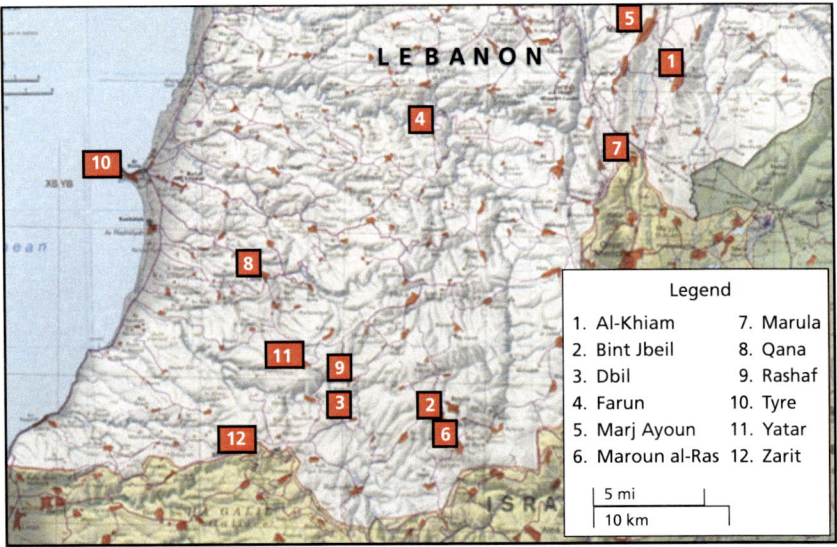

Legend

1. Al-Khiam	7. Marula
2. Bint Jbeil	8. Qana
3. Dbil	9. Rashaf
4. Farun	10. Tyre
5. Marj Ayoun	11. Yatar
6. Maroun al-Ras	12. Zarit

SOURCE: Adapted from "Southern Lebanon Border Area (1986)," courtesy of the University of Texas Libraries, The University of Texas at Austin.
RAND *MG1085-2.1*

Conflict Backdrop

Hezbollah is a Shia Islamist political and paramilitary organization that rose to prominence largely in response to Israel's occupation of southern Lebanon from 1982 to 2000. It is classified as a terrorist organization by the United States,[3] and Hezbollah and its affiliates have planned or been linked to a long series of terrorist attacks against the United States, Israel, and other Western targets, including

- "a series of kidnappings of Westerners" (including several Americans) in Lebanon in the 1980s

---

[3] Council on Foreign Relations, "Hezbollah," web page, updated July 15, 2010; U.S. Department of State, "Foreign Terrorist Organizations," web page, November 24, 2010.

- "suicide truck bombings that killed more than . . . [200 U.S. service members] at their barracks in Beirut" in 1983
- the hijacking of TWA Flight 847 in 1985
- "two major attacks" on Jewish targets in Argentina in the 1990s: the 1992 bombing of the Israeli Embassy (which killed 29) and the 1994 bombing of a Jewish community center (which killed 95).[4]

Hezbollah pursues three overall goals: the elimination of "the influence of any imperialist power" in Lebanon, Israel's "obliteration from existence and the liberation of venerable Jerusalem," and the establishment of an Islamic regime in Lebanon.[5] Hezbollah was inspired by Ayatollah Ruhollah Khomeini of Iran, and its members receive training and weapons from the Iranian Republican Guard Corps and Syria.[6]

Hezbollah advocates and practices military jihad. In the words of Sheik Naim Qassem, one of Hezbollah's founders and its current deputy secretary-general, jihad is

> another form of appraising life: death with surrender and shame versus a life of jihad ending with martyrdom for the sake of virtue's victory and national pride. In this context, Commander of the Faithful Imam Ali . . . said "Death shall defeat you in life, and you shall defeat life through death."[7]

Since its inception, Hezbollah has viewed martyrdom as a means of confronting the IDF with an asymmetric challenge. This type of martyrdom compensates for "military imbalance" by inflicting "painful losses on enemy ranks." As Secretary-General Qassem explains, these losses have, in the past, been

---

[4] Council on Foreign Relations, "Hezbollah."

[5] Augustus Richard Norton, *Hezbollah*, Princeton: Princeton University Press, 2007, p. 38–40. See also Hala Jaber, *Hezbollah: Born With a Vengeance*, New York: Columbia University Press, 1997; and Naim Qassem, *Hizbullah: The Story from Within*, Dalia Khalil, trans., London: SAQI, 2005, pp. 13–59.

[6] Council on Foreign Relations, "Hezbollah"; Jaber, *Hezbollah*, pp. 50–51, 112–113.

[7] Qassem, *Hizbullah*, pp. 336–353.

realized through simple and humble technologies that, on the one hand, shook the Israeli army's ability . . . [to] defend itself and on the other unsettled its ability to retaliate. The Israeli army withdrew in 1985 from over half of occupied territories in South Lebanon in order to reduce its spread and susceptibility to attack by the Resistance. The enemy could not tolerate many attacks like that from the pioneer of all martyr attacks, Sheikh Ahmad Kassir, who (on November 11, 1982) drove a car trapped with explosives right into the headquarters of the Israeli commander in the city of Tyre, wounding and killing 141 Israeli officers, [with] a further 10 declared missing. The enemy was forced into defeat and withdrawal from the majority of Lebanese territories on May 24, 2000, marking the largest and first liberation of its kind in the region resulting from resistance operations.[8]

In furtherance of its strategy, Hezbollah has conducted raids and suicide attacks against IDF troops and other targets in Lebanon. Hezbollah's abduction of the two soldiers in July 2006 was preceded by at least four other attempts to capture IDF personnel.[9] The organization appears to have wanted to use the soldiers as bargaining chips in an attempt to free Hezbollah personnel in Israeli custody.[10]

Hezbollah is the dominant power in southern Lebanon, and it provides education, health care, and other social services in the region. Hezbollah has developed into a political organization that holds seats in the Lebanese Parliament and wields considerable power in the country's fractured political landscape.[11]

---

[8]   Qassem, *Hizbullah*, p. 49.

[9]   Harel and Issacharoff, *34 Days*, p. 6. A few weeks before the Hezbollah ambush, Hamas militants in Gaza successfully kidnapped Gilad Shalit, another Israeli soldier, who remains in captivity as of this writing. Conversations with IDF officers make it clear that Israelis believe that it is enormously important to recover kidnapped soldiers and that the IDF is extremely careful to prevent its soldiers from being abducted.

[10]   Ina Friedman, "Moral Morass," *The Jerusalem Report*, September 4, 2006, p. 12.

[11]   For example, see "BBC News, "Who Are Hezbollah?" News.BBC.co.uk, May 21, 2008.

The State of the Israeli Military in 2006

What's in It for Us? The Problems with "Lessons Learned" from Lebanon

Before discussing how the IDF got into the condition it was in in 2006, it is important to briefly note a general shortcoming in the existing body of interpretation about why the IDF performed poorly in the Second Lebanon War. Much of this existing work approaches the subject from the perspective of determining what the "lessons" from the war "prove" about the future of warfare and what they mean in ongoing debates about what kind of military the United States and other nations will need in order to meet future challenges.[12] This existing work rarely examines Israel's decisionmaking before 2006 and therefore fails to assess Israel's perceptions about its strategic situation and its military needs. Indeed, there are some who skew interpretations of the war in parochial directions for their own purposes.

Robert Dudney's September 2006 editorial in *Air Force Magazine* is an example of the air power–centric perspective.[13] In his editorial, Dudney vigorously responds to several articles about the Second Lebanon War whose principal thesis was that the Israeli Air Force's (IAF's) "failure to halt the Shiite militia's missile attacks had 'cast doubt' on the whole 'theory' of airpower." Dudney cites several IAF contributions to the war—"IAF pilots cut Syrian and Iranian resupply routes to Hezbollah. They destroyed huge swaths of militia infrastructure. They choked off escape routes and killed hundreds of fighters. They bombed senior leadership. They supplied critical aerial reconnaissance"—and then challenges the main argument of critics of the use of air power in

---

[12] The origins of the intensified debate over the term *hybrid warfare* can largely be traced to the Second Lebanon War. See Hoffman, *Conflict in the 21st Century*, and also Stephen Biddle and Jeffrey A. Friedman, *The 2006 Lebanon Campaign and the Future of Warfare: Implications for Army and Defense Policy*, Carlisle Barracks, Pa.: U.S. Army War College Strategic Studies Institute, 2008; and Anthony Cordesman, George Sullivan, and William Sullivan, *Lessons of the 2006 Israeli-Hezbollah War*, Washington, D.C.: The Center for Strategic and International Studies, 2007. Biddle and Friedman provide a very useful list of references on pp. 2–3 that captures much of the literature on the war.

[13] Quotations in this paragraph are from Robert S. Dudney, "The Air War over Hezbollah," *Air Force Magazine*, Vol. 89, No. 9, September 2006, p. 2.

Lebanon: that Israeli airpower was not "decisive." "Precisely what," he writes, "is this supposed to mean—that IAF did not, all by itself, defeat the entrenched, highly organized Hezbollah fighters, who had six years to prepare heavily protected positions?" Dudney argues that, because of air power, "Hezbollah has been dealt a blow from which it is not likely to recover any time soon. Airpower gets a big part of the credit." Dudney also notes the potential implications for the U.S. Air Force of the argument that asserts that air power failed in Lebanon:

> These critiques of Israeli air operations are reminiscent of those that followed US Air Force successes in the 1991 Gulf War, 1995 Balkan War, 1999 Air War Over Serbia, 2001 war in Afghanistan, and 2003 war in Iraq.
>
> In those cases, some Army partisans argued that "boots on the ground," not aircraft and precision strike, contributed most to US victory. Now, as then, what is at stake are force structure, budget shares, and more.

Dudney's arguments about the contribution of air power—arguments made just a few years after the war—were articulated even more assertively several years later by Edward Luttwak, who argues that air power actually did achieve Israel's strategic objectives:

> Hezbollah leader Hassan admitted immediately after the war that he would never have ordered the original deadly attack on an Israeli border patrol had he known that Israel would retaliate with such devastating effect. Before the 2006 war, Hezbollah launched rockets into northern Israel whenever it wanted to raise tensions. Since the Aug. 14, 2006, cease-fire, Hezbollah has rigorously refrained. Whenever rockets are nonetheless launched, . . . [the] spokesmen [of Hasan Nasrallah, the secretary-general of Hezbollah] rush to announce that Hezbollah had absolutely nothing to do with it. *Evidently, Israel's supposedly futile bombing did achieve its aim.*[14]

---

[14] Edward Luttwak, "In Praise of Aerial Bombing," ForeignPolicy.com, March/April 2010. See also Martin van Creveld, "Israel's Lebanese War: A Preliminary Assessment," *The RUSI Journal*, Vol. 151, No. 5, October 2006, pp. 40–43. Van Creveld argues that,

Luttwak's article, however, has an agenda. He is using the putative outcome in Lebanon to argue for an air power–centric strategy in Afghanistan:

> The better and much cheaper alternative [in Afghanistan] would be to resurrect strategic bombing in a thoroughly new way by arming the Taliban's many enemies to the teeth and replacing U.S. troops in Afghanistan with sporadic airstrikes. Whenever the Taliban concentrate in numbers to attack, they would be bombed. This would be a most imperfect solution. But it would end the costly futility of "nation-building" in a remote and unwelcoming land.[15]

Ground power–centric arguments tend to blame Israel's problems in Lebanon on the fact that the IDF embraced an air power–centric, effects-based-operations approach to warfare. Matt Matthews's monograph, *We Were Caught Unprepared: The 2006 Hezbollah-Israeli War*, adopts this perspective and takes it even further. Matthews argues that the IDF's difficulties in Lebanon stemmed directly from the conscious neglect of Israeli ground forces by air power zealots:

> [The] Hezbollah-Israeli war was the result of a multiplicity of factors. Halutz's [Lieutenant General Dan Halutz, chief of the IDF General Staff] steadfast confidence in air power, coupled with his disdain for land warfare, increased the strength of the IAF at the expense of ground forces. While continuing COIN operations against the Palestinians, the IDF saw its budget for ground forces slashed and training for major combat operations by divisions and brigades greatly reduced. Within the IDF reserve, equipment was

---

at least for the time being, Hezbollah appears to have had the fight knocked out of it. For well over a year now, Israel's border with Lebanon has been almost totally quiet—by far the longest period of peace in four decades. This was something that neither Golda Meir, nor Yitzhak Rabin in his two terms as prime minister, nor Menahem Begin, nor Shimon Peres, nor Yitzhak Shamir, nor Benjamin Netanyahu, nor Ehud Barak, nor even the formidable Ariel Sharon, was able to achieve.

[15] Luttwak, "In Praise of Aerial Bombing."

not replaced or repaired, and the tactical skills of both reserve and regular ground forces continued to decline. Training for reserve tank crewmen was all but forgotten.[16]

Although Matthews's description of the condition of the IDF's ground forces in 2006 is largely correct, blaming their unpreparedness on General Halutz's cuts to those forces in favor of the IAF is not.

These are important points because understanding how the IDF came to be in the state it found itself in before Lebanon is central to understanding any lessons from the war, subsequent Israeli efforts to apply these lessons after the war, and the validity of these lessons for the U.S. military. The fact of the matter is that, in the aftermath of the Second Lebanon War, most Israelis did not believe that the country had achieved its strategic objectives. English historian C. V. Wedgwood's admonition is instructive in this regard: "History is lived forward but it is written in retrospect. We know the end before we consider the beginning and we can never wholly recapture what it was like to know the beginning only."[17] Again, understanding the Israelis' perceptions about their security environment before the Second Lebanon War is critical to understanding IDF performance in that conflict.

The IDF in Transition

The IDF in 2006 was in a process of conceptual transition, begun in the mid-1990s, away from a traditional, "symmetrical" view of warfare. This shift was summed up in the final report of the Winograd Commission:

> The IDF's operational concept (the old concept) was devised at the time vis-à-vis known threats in a "symmetrical" environment

---

[16] Matt M. Matthews, *We Were Caught Unprepared: The 2006 Hezbollah-Israeli War*, The Long War Series Occasional Paper No. 26, Fort Leavenworth, Kan.: U.S. Army Combined Arms Center Combat Studies Institute, 2006, p. 64. For a nuanced and useful assessment of lessons learned from the Second Lebanon War and Operation Cast Lead, see Lazar Berman, "Beyond the Basics: Looking Beyond the Conventional Wisdom Surrounding the IDF Campaigns Against Hizbullah and Hamas," SmallWarsJournal.com, April 28, 2011.

[17] C. V. Wedgwood, *William the Silent*, London: Cape, 1967, p. 35, quoted in Guenter Lewy, *America in Vietnam*, Oxford: Oxford University Press, 1978, p. 420.

that was familiar and stationary and focused on eliminating them. The "motto" included the following three elements: deterrence, early-warning, and deciding the battle. The basic idea was to concentrate a large ground force, with support from the Air Force, to transport the war quickly into enemy territory, and to attain a quick decision of the battle by capturing enemy territory and defeating the army in terms of its ability and desire to pursue the warfare. This was complemented by the basic postulation of "blue skies," namely the Air Force superiority, and the avoidance of exposing the home front to real attacks.[18]

This shift made sense for many reasons. To begin with, many believed that there had been a "transition from a pattern of symmetrical wars between regular armies and sovereign, solidified countries to asymmetrical conflicts with limited or high intensity against armed elements that rely on a sympathetic local population that assists non-government bodies from within."[19] This view had been reinforced by Israeli experiences during the two intifadas and by the low level of threat from neighboring states. A number of regional developments led to the spread of this view. First, when Israel executed Operation Defensive Shield in April 2002, reoccupying West Bank cities and besieging President of the Palestinian National Authority Yasser Arafat's compound in Ramallah in the West Bank, "the Arab world remained indifferent." This led the Israelis to conclude "that the Israeli-Palestinian conflict, severe and crisis-ridden as it might be, does not factor into the Arab states' deliberations as to launching a war against Israel, either individually or jointly."[20] The second development was the American invasion of Iraq in 2003, which led Israelis to believe that, "as long as there is a strong American presence in the region, no Arab state

---

[18] The Winograd Commission, p. 250.

[19] The Winograd Commission, p. 252.

[20] Giora Eiland, "The Decision Making Process in Israel," in Shlomo Brom and Meir Elran, eds., *The Second Lebanon War: Strategic Perspectives*, Tel Aviv: Institute for National Security Studies, 2007, p. 30.

will want to wage war against Israel."[21] These developments reinforced longstanding Israeli perceptions about its regional situation:

> Due to its clear military advantage and the peace agreements it has with Egypt and Jordan, and coupled with the lack of a basic state interest among the other Arab states—with the possible exception of Syria—in an armed conflict with Israel, since the mid 1970s Israel has enjoyed an effective and stable deterrence against all-out or even limited war vis-à-vis all the regional states.[22]

Consequently,

> the general conclusion was that since there is no entity in the Arab world interested at *present* in a war with Israel (including Hizbollah!), then a war that would erupt between Israel and one of its neighbors would result from one of two situations: either subsequent to a strategic change (a change of regime in one of the neighboring states, an American exodus from Iraq, or a change of similar magnitude), or a war launched by Israel. Common to both situations is that Israel would have strategic warning of at least several months.[23]

As a result of these estimates, defense budgets were cut significantly below the levels specified in the March 2003 multiyear defense plan known as the Kela Plan. The annual budget was eventually set at

> NIS [new Israeli shekel] 2.5 billion [approximately US$566 million] less than the Kela plan's base budget. In this situation the military rightly decided it would be correct that risk-taking be mainly in the area of war preparedness (inventory levels, technical competence, training levels). Since this area, unlike others, is

---

[21] Eiland, "The Decision Making Process in Israel," p. 30.

[22] Yair Evron, "Deterrence and Its Limitation," in Shlomo Brom and Meir Elran, eds., *The Second Lebanon War: Strategic Perspectives*, Tel Aviv: Institute for National Security Studies, 2007, pp. 36–37.

[23] Eiland, "The Decision Making Process in Israel," p. 30 (emphasis in the original).

given to changes and improvement within several months from the issue of a warning, everyone was convinced that enough lead time would be available.[24]

Israel's actual defense expenditures declined from NIS 46.9 billion (US$10.3 billion) in 2003 to NIS 44 billion (US$9.82 billion) in 2005.[25] In terms of readiness for high-intensity operations, these decisions had the greatest negative affect on the Israeli Army, particularly the reserves and armored forces (equipped with tanks and heavy armored personnel carriers). Furthermore, the ongoing costs of confronting the Second al-Aqsa Intifada (2000–2006) exacerbated the effects of the budget cuts. Indeed, the view became that "fighting is training since the intifada is the war we have to win."[26] These were rational choices based on Israeli decisionmakers' perception of the country's security situation.

The Winograd Commission notes that potential asymmetric opponents are formidable because they "are well-organized, strong, well-armed with good, effective, and modern weapons, and are equipped with advanced electronic means."[27] Furthermore, asymmetric opponents pursuing low-intensity operations present a particularly difficult

---

[24] Eiland, "The Decision Making Process in Israel," p. 30. The Winograd Commission is also very clear on this point, noting on p. 258 that

> the defense budget cuts caused the IDF, in accordance with the prioritization that it carried out in accordance with its considerations, to cut back on training and exercises, in the standing forces, and in the reserve forces alike. These constraints, together with the assessment that a significant military conflagration within the close radius to Israel (beyond the activity in the Gaza, Judea, and Samaria areas) was considered a possibility—if that—only after a gradual deterioration and escalation, or [upon] . . . Israel's initiative, led to the decision that it would be correct for the IDF to take a calculated risk and count on the option of the IDF preparing for a campaign, including gap closure in training and arming, in the event that this would be required.

[25] The International Institute for Strategic Studies, *The Military Balance 2005–2006*, London, 2005, p. 192; The International Institute for Strategic Studies, *The Military Balance 2007*, London, 2007, p. 227.

[26] Author's conversation with an IDF officer via telephone, January 17, 2011. This officer noted that these were the days when there was "no paper in the printers."

[27] The Winograd Commission, p. 252.

challenge because "the enemy can pursue the fighting for extended periods of time and the [Israelis'] ability to reach a military decision is limited." Furthermore, "capturing territory and taking control over it are not always effective to reach such a military decision, particularly in areas where the population density is high."[28] This is the situation the IDF had faced for years in the West Bank and in Gaza.

Two additional factors shaped IDF thinking about the need for new doctrinal and organizational constructs. One was the need to develop operational concepts and capabilities to deal with new types of threats. These included dealing with guerilla warfare during the occupation in Lebanon; coping with a changed Palestinian political and security situation in the aftermath of the Oslo Accords, which established an arrangement for interim self-government for Palestinians (including armed police); and managing a persistent low-intensity conflict (LIC) threat, which would spike in 1997. In late September 2000, Palestinian violence surged with the beginning of the armed Second al-Aqsa Intifada. The second factor was the IDF's decision to embrace the possibilities offered by the revolution in military affairs (RMA).[29]

Confronting Protracted Low-Intensity Conflict and Terrorism

The two intifadas created an operational need for concepts and organizations that were effective in reducing the terrorist threat to Israel. Additionally, there was the need to act within international norms by employing proportional force against terrorism while simultaneously keeping Israeli casualties low to maintain internal domestic support. The Winograd Commission notes two important considerations:

> on the one hand, the belief that the terrorist and semi-military organizations do not constitute real danger to Israel and, on the other hand, the increasing influence of the international law and

---

[28] The Winograd Commission, p. 252.

[29] Author's conversation with an IDF officer via telephone, April 12, 2011; The Winograd Commission, pp. 250–253. The IDF was confident that it had the requisite capabilities and plans to deal with Syria, its principal cross-border state threat. A looming and growing issue was what to do if Iran went nuclear.

legal rulings whose effect can limit the use of weapons and the setting of goals for assault.[30]

The first intifada (1987–1993) began as an uprising in the Jabalya refugee camp that "spread like wildfire from the Gaza Strip to the West Bank, but was largely confined to the Palestinian territories."[31] The first intifada was characterized by mass Palestinian protests that, although they were violent, primarily involved the use of "sticks and stones" rather than firearms. The IDF, however, initially responded with standing operating procedures that included the "use of live ammunition at relatively long distance from the actual source of the commotion, whenever such violence [from the protestors] was deemed life-threatening."[32] The IDF response was viewed by many as disproportional, and "the IDF became the target of ferocious criticism on the part of the Israeli Left and the international community."[33] The IDF realized that it did not have the proper tools to deal with the intifada, which was a "civilian uprising in a large-scale never before seen by the IDF." The IDF adapted, changing its rules of engagement and tactics to ones focused on riot control and LIC. It also began rethinking its concepts and doctrine.[34]

The Oslo Peace Accord brought the first intifada to a close. The accord "represented the abandonment of a Palestinian all-out struggle against Israel until its destruction as laid down in the 1968 Palestinian National Charter. It also entailed the Israeli recognition of Palestinians as an actual nation as well as their right to 'govern themselves.'"[35] The Oslo Peace Process also caused introspection within the IDF and raised several important questions, including the following: What is the role of the IDF? How would the IDF interact with the Palestinian

---

[30] The Winograd Commission, p. 253.

[31] Catignani, *Israeli Counter-Insurgency and the Intifadas*, pp. 76–77.

[32] Catignani, *Israeli Counter-Insurgency and the Intifadas*, p. 81.

[33] Catignani, *Israeli Counter-Insurgency and the Intifadas*, p. 82. Chapter Six of Catignani's book provides a detailed description of the first intifada and the IDF's adaptation.

[34] Author's conversation with an IDF officer via telephone, April 12, 2011.

[35] Catignani, *Israeli Counter-Insurgency and the Intifadas*, p. 99.

Authority police? Who is the enemy, and what do we have to achieve, when there is no opposing military to destroy, no area to occupy, and the IDF presence is itself counterproductive? What is the role of the media? Quite simply, there was a sense that something fundamental had changed and that the IDF needed different concepts to deal with a new reality. As the IDF grappled with the implications of the changing security environment, the relative quiet brought about by the Oslo Peace Accord in 1993 was shattered by an incident in late September 1996.

On September 23, 1996, Israel opened a tunnel in Jerusalem that would link the Western Wall with an exit near the Temple Mount. The Palestinian Authority responded with organized riots throughout the territories that resulted in the deaths of 14 Israelis and 56 Arabs. Armed Palestinian Authority police participated in the riots, firing on IDF soldiers. The rioting lasted some five days.[36] These riots were a turning point for Israeli decisionmakers, who abruptly realized that the Palestinians were challenging Israel with "low intensity conflict under a commitment to a political agreement, and with the threat of terror in the background."[37] That said, the Second al-Aqsa Intifada, which erupted after Ariel Sharon, then the Likud Party candidate for prime minister, visited the Temple Mount Mosque on September 28, 2000, presented the IDF with a very different situation than it had experienced during earlier confrontations with Palestinians.[38]

The greatest difference was that the Palestinians were using weapons and suicide bombers to kill Israelis. Additionally, these acts of terror were being carried out deep inside Israel in key population centers, such as Tel Aviv, Haifa, and Jerusalem. The IDF found once again that it was not fully prepared, and it was unsuccessful in stopping the large-scale violence. In March 2002, "135 Israelis were killed in 17

---

[36] Author's conversation with an IDF officer via telephone, April 20, 2011; Akiva Eldar, "Jerusalem Demolitions May Spark Repeat of 1996 Riots," Haaretz.com, last updated October 3, 2009.

[37] Central Command, Israel Defense Forces, "Washington Institute Briefing," November 1999.

[38] Catignani, *Israeli Counter-Insurgency and the Intifadas*, p. 102–103.

terrorist attacks," including 30 Israelis who died at the Park Hotel in Netanya.[39] Indeed, "Avi Dichter, then head of [Israel's General Security Service] . . . asked the Israeli public for forgiveness for the security establishment's failure to protect it."[40]

Once again, the IDF adapted. These adaptations are described in more detail later in this section; for the moment, it is sufficient to note that two factors—improved IDF operations and a security fence constructed to deny Palestinian attackers access to Israel—significantly reduced Israeli casualties. There was, however, an international backlash against IDF operations, and this reaction gave the Israeli military additional impetus to change how it operated.

International condemnation of Israeli actions became particularly strong after Operation Defensive Shield, which began on March 29, 2002. By April 3 of that year, the Israelis occupied Ramallah, Tulkram, Qalqilya, Bethlehem, Jenin, and Nablus—the six largest cities in the West Bank.[41] A report by the United Nations (UN) secretary-general notes that

> Operation Defensive Shield resulted in the widespread destruction of Palestinian private and public property. Nablus was especially hard hit, especially in its old city, which contained many buildings of cultural, religious and historic significance. Much of the destruction appears to have occurred in the fighting as a result of the use by IDF of tanks, helicopter gunships and bulldozers. . . .

---

[39] Giora Eiland, "The IDF in the Second Intifada," *Strategic Assessment*, Vol. 13, No. 3, October 2010, p. 31. See also Catignani, *Israeli Counter-Insurgency and the Intifadas*, p. 102. Catignani writes about why the Palestinians turned to high levels of violence: "[The] PA [Palestinian Authority] believed that it would be possible to achieve, through the use of violence, what *Hizbullah* had achieved, through guerrilla warfare, in Lebanon in May 2000: that is, the full unilateral withdrawal of the IDF from the Territories without the need for any formal peace agreement."

[40] Eiland, "The IDF in the Second Intifada," p. 31.

[41] United Nations, *Illegal Israeli Actions in Occupied East Jerusalem and the Rest of the Occupied Palestinian Territory: Report of the Secretary-General Prepared Pursuant to General Assembly Resolution ES-10/10*, July 30, 2002.

Much of the fighting during Operation Defensive Shield occurred in areas heavily populated by civilians and in many cases heavy weaponry was used. As a result of those practices, the populations of the cities covered in this report suffered severe hardships. The Israeli Defence Forces announced the official end of the operation on 21 April but its consequences lasted until the end of the period under review and beyond.[42]

The lessons the IDF drew from Operation Defensive Shield showed that it understood that there were gaps in its LIC capabilities and concepts. As author Sergio Catignani, describing the IDF's operations in Jenin, observed,

> Despite attempts to use precision-guided missiles and highly trained snipers to eliminate Palestinian guerrillas, whilst avoiding casualties and major collateral damage, the extensive insertion of armoured and infantry elements into such a combat battleground did bring about approximately 52-3 Palestinian deaths—most of whom were combatants—and structural damage to over 100 homes.[43]

There were significant casualties during Operation Defensive Shield. The IDF had 30 dead and 127 wounded; 240 Palestinians were killed, 500 were wounded, and more than 4,000 were detained, including 396 wanted suspects.[44] Additionally, the Israelis were sensi-

---

[42] United Nations, *Illegal Israeli Actions in Occupied East Jerusalem.*

[43] Catignani, *Israeli Counter-Insurgency and the Intifadas*, p. 167. Catignani, quoting David A. Fulghum and Robert Wall, notes that,

> as a result of such collateral damage, the IDF has, since Operation "Defensive Shield," been able to map out most Palestinian cities by developing a system that divides "the urban battlefield into precise increments and gives each building in a city . . . an individual four-digit designation so both land and air forces know exactly which target they are trying to hit." Such detailed mapping and digital designation of Palestinian urban areas has helped reduce, appreciably, the cases of operational errors, which have often led in the past to extensive collateral damage. (p. 167)

[44] Israel Ministry of Foreign Affairs, "Operation Defensive Shield: Special Update," web page, March 29, 2002; "Operation Defensive Shield (2002)," YNetNews.com, March 12,

tive to the significant international backlash to the collateral damage and civilian casualties resulting from the operation.[45]

Internally, Israeli society had changed in ways that had "practical implications . . . [for] the IDF as the army of the people."[46] These changes included

> the relative place of private goals and national goals in the citizens' program for their life; the attitude toward self-sacrifice and the need for a strong army; the shift toward a post-heroic stage in the Western culture, as part of which civilians' readiness to fight and offer their lives unless in times of a real and unequivocal existential threat has diminished considerably; the role of the media, both local and international, in the strategic environment; and the effect of the disputes among the Israeli public . . . on using military force.[47]

Consequently, Israel faced constant low-order threats of Palestinian terrorism, a relatively low existential threat environment, international pressure to exercise proportionality in its military actions, and low tolerance among its citizens for casualties and war. Understandably, the IDF began to develop operational concepts and capabilities to deal with this security environment and these internal and external constraints. This process of change, which accelerated after Operation Defensive Shield, concentrated

> on refining small-unit tactics for conducting search and arrest operations and targeted killings. Moreover, the IDF spent con-

---

2009. See also Jewish Virtual Library, "Statistics on Operation 'Defensive Shield,'" web page, undated; and United Nations, *Illegal Israeli Actions in Occupied East Jerusalem*.

[45] See United Nations Security Council, "Security Council Stresses Urgent Need for Humanitarian Access to Palestinians, Welcomes Fact-Finding Team to Examine Events at Jenin Refugee Camp," press release SC/7369, April 19, 2002. International concerns about proportionality during Operation Defensive Shield, and particularly about IDF operations in Jenin, resulted in the passage of a UN Security Council resolution and in the commissioning of a UN fact-finding mission on Jenin.

[46] The Winograd Commission, p. 252.

[47] The Winograd Commission, p. 253.

siderable time and resources in enforcing curfews and closures throughout the Territories of the West Bank and Gaza Strip. Such operations improved the IDF's constabulary and special operations capabilities.[48]

As is described below, this focus on LIC came at a cost, because it "severely impeded the IDF's training for non-urban guerrilla warfare and for preparing IDF units for large-scale joint-force operations."[49]

Thus, the IDF, and particularly the Israeli Army, was understandably—given Israeli perceptions about conditions in 2006—focused almost exclusively on LIC and on preventing the incursion of Palestinian terrorists into Israel.

Complex Concepts for Complex Problems: Systemic Operational Design

The IDF had given considerable thought to developing a theory and supporting concepts "to deal with suicide terrorists who operate from within the civilian population, a phenomenon which merited a different type of military action and preparation than what was needed in other types of conflict, including limited and asymmetrical clashes with guerrilla forces."[50] Much of the IDF's conceptual efforts to develop a new operational concept had originated in Central Command, the command that was responsible for the West Bank. The IDF had been developing LIC concepts during the first intifada in the 1990s, but the pivotal September 1996 riots reinvigorated these efforts and served as a wake-up call, alerting the IDF to the fact that that it was still involved in a low-intensity conflict and that terrorism was still a threat. In August 1997, Major General Uzi Dayan, the Central Command

---

[48] Catignani, *Israeli Counter-Insurgency and the Intifadas*, p. 190.

[49] Catignani, *Israeli Counter-Insurgency and the Intifadas*, p. 190.

[50] The Winograd Commission, p. 253. Dealing with irregular forces had been an issue since the first intifada. A former Israeli officer recounted to the author the following episode from his service as a paratrooper platoon leader during the first intifada. A group of young Palestinian schoolgirls was advancing on his platoon and throwing rocks. He made an estimate of the situation and came to the conclusion that he had three courses of action: shoot the girls, let them overrun his platoon, or run. He and his platoon ran.

commander, summed up the problem that Israel and the IDF were confronting:

> The greatest fear in the public and in the media today focuses on the outbreak of violence and terrorism . . . . [W]hat is even more bothering, however, is coping with the guerrilla-like conflict for months on end.
>
> Out of all possible scenarios it is this constant coping with low-intensity conflict (L.I.C.) for which the commanding echelon is not fully prepared.[51]

General Dayan also understood that preparing Central Command for the challenges it faced required a different way of thinking about the problem:

> The forces must be prepared for such a conflict, for only by pre-paring can we forgo a full-scale conflict. From this stems the fact that the operational field units up to the level of battalion and company will have to raise three different flags on their poles (routine security, emergency operations and war) and therefore there is the necessity to refine the way we define the terms we use. We must do so in order to create a common language which suits our field.[52]

In 1997, General Dayan began an intensive systematic effort to develop new concepts and doctrine to confront the challenges posed by the Palestinians. One of the initiatives was the creation of a Center for Low Intensity Conflict Studies. General Dayan also asked a relatively new organization that was part of the IDF General Staff to assist. This was the Operational Theory Research Institute (OTRI), which had been established in 1995 with Brigadier General (Reserve) Shimon Naveh as its director.

In May 1998, Major General Moshe ("Bogie") Ya'alon replaced General Dayan and became very personally involved in continuing his

---

[51] This material comes from slides provided to the author by an IDF officer in 2011. It quotes General Dayan.

[52] This material comes from slides provided to the author by an IDF officer in 2011. It quotes General Dayan.

predecessor's efforts. In September 2000, General Ya'alon became the IDF deputy chief of staff, and he brought many new ideas—and several key officers—to his new job. The critical issue was how to make the concepts developed for Central Command relevant for the IDF at the chief of staff level.

General Ya'alon became IDF chief of staff in July 2002, and the new concepts, collected under the term *Systemic Operational Design* (SOD), became the new IDF methodology and a key component of the 2006 IDF operational concept.[53]

The continued development of SOD was centered at OTRI. OTRI, which "was very influential in the training of the officer corps before the [Second Lebanon] war," operated on the premise "that delving into non-military post-modern theories would equip senior officers with the tools necessary for dealing with the complex and changing realities of war."[54] Indeed, "until 2003, . . . [OTRI's] core course, 'Advanced Operational Approach', was obligatory for all high-ranking Israeli officers."[55]

The ultimate expression of OTRI's approach to war, SOD, was adopted by the IDF

---

[53] One IDF officer deeply involved in the development of the new operating concept and very familiar with the evolution of SOD and IAF concepts noted that

> this was the *first* time some people [in] the IDF understood there is a difference between the tactical tools and the campaign level tools and terms. There was a need to translate the political directive, to engage with the government, to figure out what are we doing militarily in the West Bank, to consider the Israeli population, the media and so on. (Author's conversation with an IDF officer via telephone, May 13, 2010)

Additionally, as SOD concepts began to spread more broadly within the IDF during General Ya'alon's tenure, in General Headquarters and in the Israeli Army more broadly there came to be SOD "insiders" and SOD "outsiders." The outsiders felt left out (author's conversation with an IDF officer via telephone, April 12, 2011).

[54] Avi Kober, "The Israel Defense Forces in the Second Lebanon War: Why the Poor Performance?" *Journal of Strategic Studies*, Vol. 31, No. 1, February 2008, p. 31. Israeli generals maintain the title "Reserve" after leaving active duty.

[55] Eyal Weizman, *Hollow Land: Israel's Architecture of Occupation*, London: Verso, 2007, p. 187. Chapter Seven of Weizman's book contains a detailed discussion of OTRI that is based on interviews with General Naveh and others.

in response to a crisis in operational art. Symptoms of the crisis included numerous operational failures, ineffective operational thinking, and ineffective operational design. In short, SOD emerged because the traditional Israeli approach to operational design had proven ineffective in dealing with the increasing complexity of the Israeli security environment. The result was an inability to logically and purposefully bridge the gap between strategy and tactics; thus, the crisis in operational art.

Brigadier General (Reserve) Naveh and his colleagues at the Operational Theory Research Institute developed SOD after significant research into the evolution of operational art and its relation to strategy and tactics. They concluded that the traditional teleological approach to operational art based on a Western philosophy of positivism and idealism was ill suited for application in the complex Israeli security environment. As a result, they turned to emerging decision-making theories based on systems and complexity theory to develop a new approach to operational art and operational design. The result was SOD.[56]

SOD was a highly centralized process that is still controversial in the IDF and, for that matter, in the U.S. military.[57] In a November 2007 interview, General Naveh noted that SOD "is not easy to understand; my writing is not intended for ordinary mortals."[58] The interview explains the theoretical underpinnings of SOD thus:

> Naveh's art of operation is the military embodiment of system theory, an interdisciplinary theory that is used in thinking about

---

[56] L. Craig Dalton, *Systemic Operational Design: Epistemological Bump or the Way Ahead for Operational Design?* Fort Leavenworth, Kan.: School of Advanced Military Studies, U.S. Army Command and General Staff College, 2006, pp. 26–27.

[57] See Milan N. Vego, "A Case Against Systemic Operational Design," *Joint Force Quarterly*, No. 53, Second Quarter 2009, pp. 69–75.

[58] Yotam Feldman, "Dr. Naveh, or, How I Learned to Stop Worrying and Walk Through Walls," Haaretz.com, last updated October 25, 2007. General Naveh was also highly critical of the IDF leadership in this interview, stating, "The army's tragedy is that it is managed by battalion commanders who were good and generals who did not receive the tools to cope with their challenges. . . . These are people without the slightest ability in abstract thought."

computer, social and biological sciences, among others. System theory examines the operating principles of a particular unit (community, organism, computer network) through the totality of the relations between the elements that constitute it and the effect of their interactions on the overall system.

In addition to Soviet system theory, Naveh and his colleagues tried to make use of different and newer conceptual methods. He is particularly fond of the French philosophers Gilles Deleuze and Felix Guattari, authors of the books "Anti-Oedipus" and "A Thousand Plateaus." He sought to enlist their theory to describe a decentralized, irregular form of military activity, an attempt by an army to emulate guerrilla methods of operation.[59]

SOD also translated postmodernist theory into military action:

Naveh and his pupils took the Deleuze-Guattari theory, which was formulated as a philosophy of resistance and liberation and was influenced by the student revolt in France in 1968 as well as by feminist and anti-nationalist thought, and made it the theoretical underpinning for assassinations, defoliation, home demolitions and wall breaking in homes. These methods reached their peak in Operation Defensive Shield, carried out by the IDF in the West Bank in the spring of 2002.[60]

SOD was a highly centralized process that, when implemented, required detailed intelligence and real-time feedback mechanisms. SOD seems to have been useful to the IDF in understanding and operating against the Second al-Aqsa Intifada. That said, it broke down during the larger-scale, high-intensity operations in Lebanon. Furthermore, the very complex (frequently obtuse) language that characterized SOD was often confusing to many charged with its execution in Lebanon. One Israeli officer summed up what many "mortals" in the

---

[59] Feldman, "Dr. Naveh."

[60] Feldman, "Dr. Naveh." Feldman also notes that "Naveh himself established the institute in 1995 and headed it until it was dismantled 10 years later, following a harsh report by the state comptroller."

IDF thought about SOD: "It was dozens of geniuses in the General Headquarters telling everybody else what to do."[61]

The IDF and Standoff Fires

In addition to SOD, the IDF enthusiastically embraced the possibilities of the RMA, particularly the potential for improvements in capability offered by advances in ISR and precision strike. The IDF was very impressed by the U.S. performance in Kosovo in 1999, and many IAF officers had come to believe that air power alone could be a decisive force. Initial U.S. operations during the wars in Afghanistan (2001–present) and Iraq (2003–present) seemed to confirm this view, although the 2003 invasion of Iraq also caused some to begin considering the need for ground maneuver in conjunction with air operations.[62] Brigadier General Itai Brun, head of the IDF Dado Center for Interdisciplinary Studies, which replaced OTRI in 2006 after the Second Lebanon War, noted that "the Gulf War of 2003, which reflected a doctrine combining firepower and ground maneuver," affected IDF concepts more on the eve of the Second Lebanon War "than the way the United States had used air power in Kosovo."[63]

---

[61] Author's discussions with IDF officers, Tel Aviv, February 9–19, 2009.

[62] Author's conversation with an IDF officer via telephone, January 17, 2011. See Amir Eshel, "The IAF Doctrine of Counter-Terror Air Warfare," in The Fisher Brothers Institute for Air and Space Strategic Studies, *Air Power Against Terrorism, The Fisher Brothers Institute International Conference, May 2005*, Herzliya, Israel, 2005. The lecturer, an IAF brigadier general, discussed his view of the changing relationship between ground power and air power:

> Ground to air support is a most important term which needs elucidation. A ground force goes on a mission, it operates in the area and the Air Force gives support from the air. Let us think in the opposite direction. I do not suggest that the air support has been called off—if we should need it we shall be ready and able to carry it out at any time. I mean to proceed from the viewpoint that the ground force is working for the air power[.] That is[,] in the planning of a mission the force commander will do well to make a decision base[d] on the considerations of 'where will I risk less, where will I employ air power efficiently.' This is a far reaching conceptual change that the military does not quite assimilate today. . . . Air power has become more dominant even though by itself it will not be able to win the war on terror. The key [t]o success is in the jointness.

[63] Itai Brun, "The Second Lebanon War, 2006," in John Andreas Olsen, ed., *A History of Air Warfare*, Washington, D.C.: Potomac Books, 2010, p. 308.

The perceived successes of the United States in Kosovo, Afghanistan, and Iraq in using intelligence and precision air power to reach rapid decisions with few coalition casualties was extremely attractive to the IDF, particularly the IAF. Technological maturity enabled development of better IAF concepts for strategic attack, coercion, and leverage, particularly against symmetric state opponents (e.g., Syria) but also against the asymmetric LIC opponents that Israel believed it would continue to face in the future.[64]

RMA capabilities seemed to present countries capable of fielding and employing high-technology military systems—countries like the United States and Israel—with a new way of waging war. As the Winograd Commission notes, the RMA offered a "quantum leap in the development and implementation of a variety of technological abilities and counter-weapons, especially—but not only—from the air, which have the ability to identify distant targets and hit them with great precision."[65]

The overall result was a broad belief in the Israeli defense establishment that standoff fires (principally delivered by air power) were an effective means of affecting the will of the adversary and determining conflict outcomes. The standoff fires approach also seemed to promise reduced IDF casualties (a major domestic political consideration), less collateral damage (a key consideration for managing international and regional opinion), and budgetary savings.

The standoff fires approach, however, is uniquely Israeli, and it differs from the U.S. Air Force's concept of effects-based operations (EBO). The Israeli approach relies on a more quantitatively based targeting process focused on attaining "desired achievement"—rather than "effects"—against adversary targets. In the Israeli approach, standoff fires are used to attack adversary systems and to enable the achievement of desired operational outcomes.[66] The EBO concept is outlined in the following paragraph from U.S. Air Force doctrine:

---

[64] Author's conversation with an IDF officer via telephone, April 12, 2011.

[65] The Winograd Commission, p. 252.

[66] Author's conversation with an IDF officer via telephone, January 17, 2011.

Effects-based actions or operations are those designed to produce distinct, desired effects while avoiding unintended or undesired effects. This new conceptual model requires that airmen think through the full range of outcomes, choose those that will best achieve objectives, and find ways to mitigate those that will impede achieving them. Air and space power offers many different ways to achieve a given effect; the effort of thinking through actions in this manner should yield commanders and national leaders many options beyond attrition or annihilation. Therefore, adoption of EBO also requires that airmen advocate air and space power's capabilities in terms of desired effects rather than targets.[67]

Thus, although the IDF (particularly IAF) standoff fires and U.S. Air Force EBO concepts both rely on RMA technologies for radical improvements in capabilities and for achieving operational and strategic ends, there is a not-so-subtle difference between them: The Israeli approach is more quantitative and target-centric, and it focuses on the efficient identification and destruction of specific targets, assuming that these actions will yield success. The U.S. Air Force doctrine is decidedly more qualitative and focused on "desired effects."[68]

In the words of one IAF officer, the Israeli approach is "effects-oriented" rather than "effects-based" and is focused on "specific outcomes—what do we want to achieve?" This is not to say, however, that the "achievement" does not transcend the specific target. For example, this officer noted that there are two ways of taking down a building: "destroying each brick or destroying the foundation; it is a question of what is the center of gravity to cause enemy collapse; what

---

[67] U.S. Department of the Air Force, AFDD 1, *Air Force Basic Doctrine*, AFDD 1, Washington, D.C., 2003, p. 18. For a discussion of the origins of EBO in the United States, see David E. Johnson, *Learning Large Lessons: The Evolving Roles of Ground Power and Air Power in the Post-Cold War Era*, Santa Monica, Calif.: RAND Corporation, MG-405-1-AF, 2007, pp. 186–189.

[68] As one IDF officer noted, "The purpose of the Air Force is to hit targets; the purpose of the Army is to identify targets" (author's discussions with IDF officers, Tel Aviv, March 2–5, 2008).

are the right targets to bring the system down."[69] In short, specific outcomes are based on targets.[70]

SOD concepts (which involve many of the systems-oriented tenets of EBO) and the idea of standoff fires were merged into an Israeli operational concept that was accepted by General Halutz in April 2006.[71] An article by Alon Ben-David notes that the doctrine provided a

> "conceptual framework" for military thinking, replacing traditional notions of "objective" and "subjection" with new concepts like "campaign rationale" and "conscious-burning" of the enemy. The doctrine's aim was to recognise the rationale of the opponent system and create an "effects-based" campaign consisting of a series of "physical and cognitive appearances" designed to influence the consciousness of the enemy rather than destroying it.[72]

The language in this article seems to have originated in the more-qualitative language of SOD rather than in the more-quantitative concepts of standoff fires. Indeed, as Dima Adamsky notes,

---

[69] Author's conversation with an IDF officer via telephone, April 12, 2011. A good example of this is targeting specific leaders to remove them and to keep the remaining leaders on the run all the time.

[70] Author's conversation with an IDF officer via telephone, April 12, 2011. This officer noted that, although the initial concept within the IAF was EBO, it was replaced by effects-oriented operations. "Effects" had been dropped from IAF operational language and replaced with "desired achievement" before 2006 because the term *effects* was "vague and unclear." The term *effect* was never used in IAF plans or orders during the Second Lebanon War.

[71] The Winograd Commission refers to this concept as the "Anthology of the General Staff Command's Operational Concept for the IDF, April 2006, 1st Edition" (p. 262). General Halutz replaced General Ya'alon as IDF chief of staff in June 2005. See also Itai Brun, "The Second Lebanon War, 2006," unpublished manuscript, undated [2009], p. 31. General Brun notes the following about the new operational concept:

> It is unlikely that readers had time to give it serious consideration before the [Second Lebanon] war began. In fact, it did not even represent a complete doctrine. Its importance stems from the fact that it reflects the way the senior military leadership interpreted the series of changes in both the Israeli strategic environment and Israeli society itself and how those changes influenced IDF doctrine in the relevant period.

[72] Alon Ben-David, "Debriefing Teams Brand IDF Doctrine 'Completely Wrong,'" *Jane's Defence Weekly*, January 3, 2007.

the new theory of victory strongly contrasted with classical linear operations, where the enemy was brought to its knees in a decisive battle of annihilation. The new approach demanded simultaneous attack throughout the entire depth and dimensions of operational deployment that would create overall paralysis of the enemy system. This type of victory not only was based on physical annihilation, but paid a great deal of attention to affecting the rationale of an enemy system and paralyzing its motivation and ability [to] keep on fighting. . . . [T]his new theory of victory relied heavily upon the American concept of Effect-Based Operations (EBO). EBO advocated applying military levers not only for the sake of inflicting damage per se, but in order to produce indirect and cascading effects that would influence the enemy as a system and attain the strategic goals of the campaign.[73]

A key element of the new concept, according to the Winograd Commission, "was a change in the attitude toward the critical importance of a large-scale ground exercise as an element that helps determine the military decision."[74] This view was fostered by a belief in

the cumulative result of technological changes that enabled high-power precision firing and of confrontation conditions that precluded the option of concentrating a force in a simple way and

---

[73] Dima Adamsky, *The Culture of Military Innovation: The Impact of Cultural Factors on the Revolution in Military Affairs in Russia, the US, and Israel*, Stanford, Calif.: Stanford University Press, 2010, p. 106. As one IDF insider noted,

the "new operating concept" was a unique Israeli conceptual framework, combining SOD, [the] effects approach, air-centric [operations], [the] LIC mindset, and [the] RMA—tailored to the Israeli context. Its main breeding ground was the Central Command because there they have realized that regular military TTPs [tactics, techniques, and procedures] and terms are not relevant. . . . This was the *first* time some people [in] the IDF understood there is a difference between the tactical tools and the campaign-level tools and terms. There was a need to translate the political directive, to engage with the government, to figure out what are we doing militarily in the West Bank, to consider the Israeli population, the media, and so on. (Author's discussion with an IDF officer, May 13, 2010)

[74] The Winograd Commission, p. 255.

of attaining a clear military decision by conquering an area and destroying the adversary's military capabilities.[75]

This view had tremendous implications because it changed "the purpose of the firing from being a supportive element to serving as a central component in attaining the decision."[76] This construct, which Israeli analyst Ron Tira calls "standoff fire operations,"[77] seemed to promise the ability to achieve strategic and operational goals without requiring the following:

a. A deep, large-scale land maneuver to advance firing weapons for the purpose of attacking targets and achieving effects deep inside the territory;

b. Massive takeover of enemy territory in order to prepare the space that the large assault echelons needed to conduct battle-deciding mobile battles;

c. The need to conquer large areas to have control over them or to use them as a "bargaining chip" to create pressure and reach political negotiations from an advantage[ous] position.[78]

The Winograd Commission sums up the essence of the operational concept thus: "Identifying the aerial medium and its superiority as a central element enables more effective utilization of the maneuver, collection, destruction, and control capabilities . . . while minimizing friction opposite the asymmetrical elements that the enemy had developed."[79]

---

[75] The Winograd Commission, p. 255.

[76] The Winograd Commission, p. 255.

[77] Ron Tira, *The Limitations of Standoff Firepower–Based Operations: On Standoff Warfare, Maneuvers, and Decision*, Tel Aviv: Institute for National Security Studies, 2007, p. 13.

[78] The Winograd Commission, p. 255.

[79] The Winograd Commission, p. 255.

Systemic Operational Design and Standoff Fires in Action: Defeating the Intifada

The SOD and standoff fires operational concepts appeared to be validated during the Second al-Aqsa Intifada.[80] A central component of Israeli operations against Palestinian terror activities became known as "mowing the grass," a euphemism for killing or capturing Palestinians to disrupt their ability to carry out acts of terror in Israel. These operations involved all relevant elements of the Israeli intelligence and security forces in finding and engaging fleeting targets who were continually adapting to Israeli efforts.[81] Technical intelligence from sensors, UAVs, and human sources was provided rapidly to Israeli operational units via improved command, control, and communications systems. Human intelligence was also key: The IDF used *"Mistar'aravim* (i.e. 'to become an Arab') units, who are trained to blend in with the local population, gather operational intelligence and sometimes conduct targeted killings."[82] The Israeli General Security Service also employed Palestinian collaborators for intelligence collection, "particularly in the targeted killing of key terrorist leaders."[83] In short, the IDF developed processes across the government that provided exquisite intelligence to support the fight against the intifada. Importantly, the Israelis were able to concentrate all of the necessary resources on the specific operation at hand. This would not be the case in the Second Lebanon War, when there were so many actions occurring simultaneously that the capacity of the intelligence system to respond was exceeded.[84]

Sensor-to-approval-to-shooter decisions were very time sensitive and were heavily influenced by the imperative to minimize civilian

---

[80] Adamsky, *The Culture of Military Innovation*, p. 107.

[81] Author's discussions with IDF officers, Tel Aviv, March 2–5, 2008, and February 9–19, 2009. "Mowing the grass" operations focused on continually disrupting terrorist networks by killing or capturing leaders. The expectation was that these actions would degrade the networks by inducing turbulence that would prevent them from recovering and from attaining their original levels of competence.

[82] Catignani, *Israeli Counter-Insurgency and the Intifadas*, p. 113.

[83] Catignani, *Israeli Counter-Insurgency and the Intifadas*, p. 113.

[84] Author's conversation with an IDF officer via telephone, April 12, 2011.

casualties and collateral damage. Who had to approve the attack of a target depended on the target's political significance, and the approving authority ranged from local commanders to the Israeli prime minister. Over time, closely integrated small-unit operations with highly centralized control evolved, and these operations employed standoff fires (generally provided by attack helicopters) to attack targets. These operations were also discretionary and highly contingent, with targets being engaged only if the prospect of success was high and the risk of Israeli casualties, unintended civilian casualties, and unintended collateral damage was low.[85] Importantly, *there was no need to accomplish any specific operation rapidly.*[86] Indeed, the imperatives became "zero casualties to our forces"[87] and "zero malfunctions";[88] this was especially true after the IDF suffered a fairly high number of casualties during Operation Defensive Shield.[89]

The effectiveness of these concepts during the Second al-Aqsa Intifada, if one's measure is reductions in Israeli casualties, was impressive: In 2002, 2,348 Israeli civilians and security forces were wounded and 220 died; in 2005, 660 were wounded and 22 died.[90] Nevertheless, within the IDF there began to emerge a pattern that one IAF officer characterized as a "time-management problem." Getting things absolutely right and avoiding casualties became more important than rapid mission accomplishment. In short, the IDF's collective sense of urgency eroded during these discretionary operations.[91]

Finally, the nature of operations against the intifada—operations that combined centralized, discretionary operations with high-

---

[85] Author's discussions with IDF officers, Tel Aviv, March 2–5, 2008, and February 9–19, 2009.

[86] Author's conversation with an IDF officer via telephone, January 17, 2011.

[87] Yehuda Wegman, "The Struggle for Situation Awareness in the IDF," *Strategic Assessment*, Vol. 10, No. 4, February 2008, p. 23.

[88] Author's conversation with an IDF officer via telephone, April 13, 2011.

[89] "Operation Defensive Shield (2002)."

[90] Israel Ministry of Foreign Affairs, "Victims of Palestinian Violence and Terrorism Since September 2000," web page, undated.

[91] Author's conversation with an IDF officer via telephone, January 17, 2011.

technology sensors and real-time decisionmaking—had, by the time of the Second Lebanon War, become heavily tinged by what the Israelis later called the "plasma-screen effect." Commanders became tied to their headquarters, where they watched unmanned aerial vehicle (UAV) feeds on plasma screens and directed the activities of their subordinates from rear areas rather than getting close to the action with forward units.[92] The Winograd Commission notes that, during the Second Lebanon War, at "the division and formation commanders' levels, . . . we found that some of the commanders ran the fighting from inside Israeli territory, not even on the border itself, but from the Northern Command or another command center."[93]

The Israeli Elite's Misperceptions About the Security Environment

The absence of a high-end threat to Israel, combined with the significant demands on the IDF to deal with the Second al-Aqsa Intifada (and with the IDF's forced removal of Israeli settlers from Gaza in 2005), significantly affected the strategic and operational perspectives of Israeli politicians and military officers. The result was a LIC-centric mindset:

> Some of the political and military elites in Israel have reached the conclusion that Israel is beyond the era of wars. It had enough military might and superiority to deter others from declaring war against her; these would also be sufficient to send a painful reminder to anyone who seemed to be undeterred; since Israel did not intend to initiate a war, the conclusion was that the main

---

[92] Author's discussions with IDF officers, Tel Aviv, March 2–5, 2008, and February 9–19, 2009. See Eitan Shamir, *Transforming Command: The Pursuit of Mission Command in the U.S., British, and Israeli Armies*, Stanford: Stanford University Press, 2011. Shamir notes that

the postwar account of the Ninety-first Division identified a clear preference for control over command. Orders lacked clear statements of purpose or critical constraints such as time. In general, initiative, once the hallmark of the IDF, was exercised only by the lowest echelons. Command from the front was rarely practiced, even in dire straits. (p. 153)

[93] The Winograd Commission, p. 385.

challenge facing the land forces would be low intensity asymmetrical conflicts.[94]

The Israeli Army was the service most affected by this view of the security situation, particularly because LIC requirements had stretched it significantly and budget cuts had been implemented. Furthermore, because the Israeli Army believed its future demands were going to involve LIC, it understandably focused preparing for this type of operation. Consequently, high-end combat training was not deemed important. The high operational tempo and reduced training funds combined to produce a de facto view that fighting the intifada was all the training needed by ground units—the view that "fighting is training." Additionally, because active-duty Israeli Army units were principally responsible for ongoing LIC operations, which were mainly carried out by brigade formations, reserve training and higher-level (e.g., division-level) exercises were curtailed.[95]

On the eve of combat operations in Lebanon in 2006, the Israeli Army was deeply conditioned by its protracted LIC experiences, which "created a misconception of what war is really like," particularly since "at no stage [during the Second al-Aqsa Intifada] was an Israeli unit required to face down an enemy force of a size larger than an unskilled infantry squad."[96] Armored units (i.e., tanks and mechanized infantry) played only a small role in these operations. Armored-unit training was neglected because those units were deemed largely irrelevant in LIC. Finally, the Israeli Army assigned its best officers to deal with LIC threats in the West Bank and Gaza, further reinforcing the focus on LIC:

> Service in the West bank and Gaza strip became mandatory for career advancement. . . . The situation on the Lebanese border was less auspicious, even for those at the level of staff officers. In

---

[94] Israel Ministry of Foreign Affairs, "Winograd Commission Submits Interim Report," web page, April 30, 2007.

[95] Author's discussions with IDF officers, Tel Aviv, March 2–5, 2008, and February 9–19, 2009.

[96] Harel and Issacharoff, *34 Days*, p. 45.

general, the best commanders were assigned to the occupied territories, not to Lebanon[97]

The view that LIC was the future was also reflected in perceptions about the utility of weapons. One IDF officer assigned to the General Staff recounted that infantry units were asking that they be allowed to warehouse their mortars and other heavy weapons because they never used them. Israeli Army units that went into Lebanon came to find that these weapons, which they had not themselves been trained to use, were essential in fighting Hezbollah. In the Second Lebanon War, the Israeli Army would pay a high price in lives and prestige for the fact that its conventional, high-end warfighting skills had been allowed to atrophy.[98]

Like the Israeli Army, the IAF was heavily engaged in fighting the Second al-Aqsa Intifada and in confronting the challenges of LIC, but it was also the principal means of deterring and hedging against the unlikely possibility of a state-on-state war and of carrying out possible strikes against Iranian nuclear capabilities. Thus, it had to maintain a mix of low- and high-end capabilities.

For LIC, the IAF had made significant adaptations in its "munitions . . . [and] command and control and intelligence systems, and [had adopted] new methods of air operations."[99] In an article based on an interview with an IAF colonel, journalist B. C. Kessner reports that, in the area of munitions,

> one of the IAF's first realizations was that it had to control the intensity of its munitions. This required a shift from maximum

---

[97] Harel and Issacharoff, *34 Days*, p. 63.

[98] Author's discussions with IDF officers, Tel Aviv, March 2–5, 2008, February 9–19, 2009, and September 2–10, 2009; Gabriel Siboni, "The Military Campaign in Lebanon," in Shlomo Brom and Meir Elran, eds., *The Second Lebanon War: Strategic Perspectives*, Tel Aviv: Institute for National Security Studies, 2007, p. 66. For excellent examinations of the state of the IDF prior to the Second Lebanon War, see Shlomo Brom and and Meir Elran, eds., *The Second Lebanon War: Strategic Perspectives*, Tel Aviv: Institute for National Security, 2007; Catignani, *Israeli Counter-Insurgency and the Intifadas*; and Harel and Issacharoff, *34 Days*.

[99] B. C. Kessner, "New Department Transforming IAF for LIC Success," *Defense Daily International*, Vol. 6, No. 10, March 11, 2005, p. 1.

lethality to maximum precision and accuracy, for targets like individual cars and buildings. "We have made very significant steps," the colonel said. He described recent actual attacks where only the right room in a building was hit through a window, not affecting the other rooms, and others where only a floor and not the whole building, or only a car and not others surrounding it, were hit by LIC-generation weapons.[100]

The interviewed officer emphasized that "the goal right from the beginning was to be able to hit the target, any target, and only the target . . . [minimizing] things like collateral damage and the killing of innocent civilians."[101] As in the case of "mowing the grass operations," precision strikes and near-real-time integration of intelligence (often from multiple sources) were required.[102]

In the IDF, almost all aviation assets are owned by the IAF, including UAVs, attack and utility helicopters (including those used for medical evacuation), and fixed-wing aircraft.[103] By 2006, the IAF had become quite adept at supporting the Israeli Army in LIC contexts, particularly in using attack helicopters to "pinpoint and eliminate hostile forces by using snipers or missile attacks."[104] Nevertheless, the IAF's capacity to support large-scale air-ground operations—particularly in the area of close air support (mostly fixed-wing close air support)—was neglected by both the IAF and the Israeli Army. Indeed, air-control parties were removed from Israeli Army brigades and there was little, if any, joint training.[105] Again, from the point of view of those making decisions before the Second Lebanon War, such training was not neces-

---

[100] Kessner, "New Department Transforming IAF for LIC Success," p. 1.

[101] Kessner, "New Department Transforming IAF for LIC Success," p. 1.

[102] Kessner, "New Department Transforming IAF for LIC Success," p. 1.

[103] Author's discussions with IDF officers, Tel Aviv, March 2–5, 2008, and February 9–19, 2009. See International Institute for Strategic Studies, *The Military Balance 2009*, Milton Park, UK: Routledge, 2009, p. 250, which lists the types and numbers of aircraft in the IAF.

[104] Catignani, *Israeli Counter-Insurgency and the Intifadas*, p. 113.

[105] Author's discussions with IDF officers, Tel Aviv, March 2–5, 2008, and February 9–19, 2009.

sary, given the absence of a threat that would require that the IDF be ready to employ this kind of capability.

IDF Deficiencies on the Eve of the Second Lebanon War

When the IDF was about to go to war in Lebanon, it had a number of key deficiencies:

- It was in the early stages of incorporating a new operational concept that was highly reliant on SOD and standoff fires (provided by both air and artillery) to change the opponent's will. There was no clear understanding of whether this approach would have a high probability of success against such an adversary as Hezbollah.[106]
- It did not understand Hezbollah sufficiently. Intelligence about Hezbollah was highly compartmentalized and not generally available to operational commanders and planners.[107]
- The Israeli Army was largely unprepared to carry out high-intensity, combined-arms fire-and-maneuver operations.
- The IAF and the Israeli Army had not trained or organized for joint air-ground operations.
- The exquisite intelligence that the IDF had become accustomed to having in the West Bank and Gaza was not available in Lebanon.[108]
- Neither the IAF nor the Israeli Army had a clear understanding of how to address the persistent short-range rocket attacks that Hezbollah launched against Israel once the war commenced.[109]

---

[106] Author's discussions with IDF officers, Tel Aviv, March 2–5, 2008, February 9–19, 2009, and September 2–10, 2009.

[107] Author's conversation with an IDF officer via telephone, January 17, 2011.

[108] Yoaz Hendel, "Failed Tactical Intelligence in the Lebanon War," *Strategic Assessment*, Vol. 9, No. 3, November 2006.

[109] Author's discussions with IDF officers, Tel Aviv, March 2–5, 2008, February 9–19, 2009, and September 2–10, 2009. See Harel and Issacharoff, *34 Days*, pp. 86–90. There seems to have been a sense among the majority of Israeli decisionmakers that the initial retaliatory air strikes against Hezbollah and limited bombing of Lebanese infrastructure (e.g., the Beirut airport tarmac, the Beirut-Damascus highway) would decide the issue. Harel and Issacharoff

Additionally, Northern Command, the responsible headquarters during the war, had no operational plans for dealing with these attacks once they began.[110]

- Israeli political and military leaders did not understand Hezbollah's strategy of "not losing is winning."[111]

It is also important to understand that the IDF's operational concept was not yet thoroughly institutionalized. Indeed, as Dima Adamsky observes,

> when Israel surprised itself with the decision to go to war, the IDF was experiencing the climax of its conceptual disorientation. . . . This bureaucratic-conceptual chaos and doctrinal relativism were among the factors that made the IDF war machine somewhat dysfunctional during the war.[112]

All of these issues, as will soon be seen, would cause problems for Israel in Lebanon.

The Challenges Posed by the Terrain and by Hezbollah

Hezbollah had been preparing for a conflict with Israel ever since the Israeli withdrawal from southern Lebanon in May 2000. In the Second Lebanon War, the IDF confronted challenges presented both by the complex natural terrain in Lebanon and by Hezbollah's preparation of

---

quote a senior Israeli minister who believed that "the Israeli government decided not to go to war, but to launch an operation that would inform Nasrallah loud and clear—that the time of abductions was over" (p. 85). The IDF actions triggered Hezbollah rocket attacks—a reaction anticipated by a small number of Israeli ministers and military officers. Harel and Issacharoff note that "only later did they [the ministers] realize that the IDF had no plans for defeating Hezbollah or countering short-range Katyusha fire" (p. 87).

[110] Author's conversation with an IDF officer via telephone, January 17, 2011.

[111] Author's conversation with an IDF officer via telephone, January 17, 2011.

[112] Adamsky, *The Culture of Military Innovation*, p. 109.

the battlefield. Additionally, Hezbollah was a well-armed and a well-prepared fighting force.

Terrain

The terrain of southern Lebanon offers decided advantages to the defender. It consists of both lowlands and highlands, and stark variations in climate, soil type, and vegetation occur across short distances. Tyre is the only major city in the region, and many of the villages, where most of the predominantly Shia population live, are located on top of small hills separated by deep valleys. The nature of the landscape restricts and channelizes armored maneuver and forces wheeled vehicles to remain on easily identifiable and exposed roads that are ideal sites for ambushes and for the employment of mines and IEDs. The terrain demands well-trained infantry and integrated fire and maneuver.[113] Although armored forces (tanks and armored personnel carriers) faced mobility challenges in the terrain of southern Lebanon, they can, if employed properly, provide vital protected mobility and firepower.

Hezbollah's Preparation of the Battlefield

Hezbollah took advantage of the difficult terrain in southern Lebanon in developing its defenses. In 2000, it began to build a network of bunkers, trenches, tunnels, and fighting positions across southern Lebanon, both in the villages and in the "nature reserves."[114] The bunkers were often quite sophisticated and well equipped with electricity, phone lines, air conditioning, and stores of food, water, weapons, and ammunition. Most had a sleeping area, a kitchen, and a bathroom. It is suspected that Hezbollah built the structures with Iranian, and possibly North Korean, help.[115] Some bunkers were 20–30 meters under-

---

[113] Andrew Exum, *Hizballah at War: A Military Assessment*, Policy Focus No. 63, Washington, D.C.: The Washington Institute for Near East Policy, 2006, pp. 2–3.

[114] Author's email exchange with an IDF officer, December 10, 2010. "Nature reserve" was the IDF's code name for fortified rocket-launching sites hidden in bushy terrain.

[115] Matt M. Matthews, "Hard Lessons Learned: A Comparison of the 2006 Hezbollah-Israeli War and Operation CAST LEAD: A Historical Overview," in Scott C. Farquhar, ed., *Back to Basics: A Study of the Second Lebanon War and Operation CAST LEAD*, Fort Leavenworth, Kan.: Combat Studies Institute Press, 2009, p. 9. See also Exum, *Hizballah at*

ground and featured rocket-firing positions. Some were "constructed in great secrecy" along the Israeli-Lebanese border, with their "entrances cunningly camouflaged." The walls and ceilings were often "reinforced with steel plates and girders painted black to prevent stray reflections from the sun from giving away the concealed entrance."[116] The hilltop villages in southern Lebanon were often "de facto fortresses" that offered cover, concealment, and clear fields of fire to Hezbollah defenders.[117]

Hezbollah: A Hybrid Adversary

Hezbollah did not just take advantage of the terrain and prepare the battlefield where it would fight; it also used its hybrid nature to challenge the IDF. Hezbollah's leaders came to the conclusion that, although Hezbollah possesses some capabilities associated with conventional military forces (e.g., effective standoff weapons), the organization would be best served by focusing fighter training on unconventional means, with an emphasis on fighters operating individually and in small groups.[118] Hezbollah trained its forces in Lebanon, Iran, and Syria and emphasized creating units that combined the weapons normally associated with states with guerilla warfare tactics (e.g., ambushes, attack and withdraw)—a combination that is the essence of hybrid warfare.[119]

War, p. 3. As Exum notes, "For the defender, the rocky, hilly landscape of southern Lebanon provides several advantages. The terrain, while allowing unrestricted dismounted movement, largely restricts armored maneuver, channeling vehicles toward roads or other easily identifiable avenues of approach."

[116] Nicholas Blanford, "A Rare Trip Through Hizbullah's Secret Tunnel Network," CSMonitor.com, May 11, 2007.

[117] Exum, *Hizballah at War*, p. 3.

[118] Based on a RAND colleague's discussion with a Hezbollah military analyst, Beirut, January 2009.

[119] Based on a RAND colleague's discussion with a Lebanese military analyst, Beirut, January 2009. Hezbollah has three distinct training methods for different types of conflict: unconventional and militia warfare training conducted in Lebanon, Syria, and, to some extent, Iran; "street fighting" training conducted mainly in Lebanon; and "conventional army training" conducted mainly in Iran.

In 2006, the Israelis estimated that Hezbollah had 10,000 fighters. Although it was active throughout Lebanon, Hezbollah was concentrated in southern Lebanon, Beirut, and Baalbek. Hezbollah's headquarters (and Secretary-General Nasrallah's offices) were located in the Dahiye neighborhood of the Shia section of Beirut. The majority of Hezbollah's fighters were located in the Nabatieh region and south of the Litani River.[120] Although exact numbers are difficult to ascertain, published sources state that Hezbollah relied almost exclusively on the 3,000 fighters in the Nasr Brigade and that "Hezbollah never felt the need to reinforce" that brigade.[121]

As Nicholas Blanford describes, Hezbollah was organized in a cellular manner in southern Lebanon:

> The IR [Islamic Resistance, Hezbollah's military wing] splits south Lebanon into several sectors, each one consisting of between 12 and 15 villages. Each sector was subdivided into smaller components of two to three villages. All sectors remained in close contact with each other and the IR leadership in Beirut using sophisticated fibre-optic communications that resisted IDF jamming and interception measures. A Hizbullah official said that each sector had the power to act autonomously if communications were severed, although IR's chain of command remained unbroken throughout the conflict. On the sub-sector level, fighters used Motorola walkie-talkies. Each fighter was issued a code number and communicated using an ad hoc code based on local and personal knowledge of each other that would be meaningless to an eavesdropper.
>
> The IR forces on the ground in south Lebanon were split essentially into two wings. The first was the full-time military force of experienced, well-trained, highly disciplined and motivated

---

[120]Brun, "The Second Lebanon War," pp. 8–9.

[121]Alastair Crooke and Mark Perry, "How Hezbollah Defeated Israel: Part 2—Winning the Ground War," ATimes.com, October 13, 2006. See also Yaakov Katz, "IDF Declassifies Intelligence on Hizbullah's Southern Lebanon Deployment," *Jane's Defence Weekly*, July 9, 2010. Katz's article estimates that, as of July 2010, Hezbollah had "30,000 guerrilla fighters—20,000 deployed in southern Lebanon—compared with 15,000 in 2006."

guerrilla fighters, aged from their late twenties to late thirties. Numbering a few hundred, the full-timers were deployed in the network of bunkers and tunnels in south Lebanon as well as other locations. These fighters, equipped with military uniforms, were split into teams of 15 to 20 and chiefly were responsible for artillery rockets, advanced anti-tank missiles and sniping. . . .

The second wing was the "village guard" units, many of them veteran guerrilla combatants from the 1990s when the IDF occupied south Lebanon. Although they share the same high degree of motivation and discipline as their full-time comrades, the village guards were an irregular force of part-time personnel. The guards remained in their villages after most civilians had fled north. In the event of an IDF ground invasion, the village guards would provide successive layers of defence consisting of fresh, well-armed fighters able to take advantage of their intimate knowledge of the local terrain to interdict and frustrate the IDF advance. The village guards, dressed in civilian clothing, were armed with AK-47 assault rifles, light machine guns, rocket-propelled grenades and AT-3 Sagger anti-tank missiles.[122]

In southern Lebanon, most Hezbollah fighters are locally based, fighting near their homes. They live with their families and lead a normal life. When called up, a fighter reports to his unit. There is no fixed unit size, but, depending on the area of operation, units in these local fighting units can have between three and 15 members. Because the fighters live near their operating areas and rely mainly on stockpiled supplies, there is no need to provide transportation or to protect lines of communication. This decentralization helps Hezbollah conduct operations that are less vulnerable to Israeli interdiction.[123] Finally, Hezbollah fighters are very familiar with their area of opera-

---

[122]Blanford, "Deconstructing Hizbullah's Surprise Military Prowess."

[123]Based on a RAND colleague's discussion with a Hezbollah military analyst, Beirut, January 2009.

tions, enjoy widespread popular support, and have very effective communications networks.[124]

In the southern suburbs of Beirut and in the Bekaa Valley, most Hezbollah fighters expected to operate near their homes, but, if necessary, they could have been mobilized and sent to the south. In the event of a war, they were to keep in close communication with the central command and receive direct orders, but, if communications were disrupted, they were to retain the ability to operate and to fight without referring to their commands.[125]

During the 2006 war, Hezbollah fighters sometimes wore Israeli military uniforms to deceive and confuse the Israelis. Using Hezbollah's extensive network of trenches, tunnels, and bunkers for cover and for stockpiling weapons and supplies, Hezbollah fighters waited for Israeli patrols to approach. Once IDF soldiers advanced inside a village, Hezbollah personnel used surprise and close proximity to inflict damage on the Israelis. Often, one Hezbollah unit focused on fighting IDF infantry while another sought to destroy or disable tanks.[126]

Away from the border, particularly near main roads, other Hezbollah fighters with extensive training in weapon use (including ATGMs) sought to disable or destroy IDF HMMWVs and tanks from standoff distances. They had to travel to their fighting areas, but their weapons were cached near where they expected to operate. These fighters sought

---

[124]Jeremy M. Sharp, Christopher Blanchard, Kenneth Katzman, Carol Migdalovitz, Alfred Prados, Paul Gallis, Dianne Rennack, John Rollins, Steve Bowman, and Connie Veillette, *Lebanon: The Israel-Hamas-Hezbollah Conflict*, Washington, D.C.: Congressional Research Service, Library of Congress, 2006, p. CRS-10.

[125]Based on a RAND colleague's discussion with a Hezbollah military analyst, Beirut, January 2009.

[126]Author's discussions with IDF officers, Tel Aviv, February 10–12, 2009; author's discussions with IDF officers, Washington, D.C., February 26, 2009; author's discussions with IDF officers, Tel Aviv, April 1, 2009, and September 2–10, 2009; Arwa Mahmoud, "Kital Hezbollah, al-Din fi mouwajahat Isra'il, kifa intasar Hezbollah fi harb tamouz 2006 [Hezbollah's Fight, Religion in the Confrontation with Israel, How Hezbollah Won in the July 2006 War]," Amir Publication, 2008, pp. 143–145.

to slow IDF maneuver and to cause casualties using surprise and hit-and-run tactics.[127]

It is unknown how many ATGMs Hezbollah had in southern Lebanon. One estimate is that it had an inventory of "several thousand ATGMs," ranging from "elderly" Saggers to very modern Kornets.[128] The method of employment described by defense analyst David Eshel indicates that Hezbollah had significant numbers of ATGMs:

> Hezbollah deployed their tank-killer teams in a thin but effective defensive scheme, protecting the villages where the organization's Shiite members reside; villages where their short range rockets were positioned and where command infrastructure and logistics support was set up. An estimated 500 to 600 members of their roughly 4,000-strong Hezbollah fighting strength in South Lebanon were divided into tank-killer teams of 5 or 6, each armed with 5–8 anti-tank missiles, with further supplies stored in small fortified well camouflaged bunkers and fortified basements, built to withstand Israeli air attacks.[129]

Hezbollah also stockpiled a rocket force to give it the capability to conduct indirect-fire missions into Israel. It has declared that it sought to keep a regular and consistent rate of fire from its rocket force into Israel. Hezbollah personnel who worked with rockets and missiles received specialized training and operated in areas where they could hide from the IAF and where Israeli infantry could not move freely.[130] Table 2.1 shows the broad array of Hezbollah rockets in Lebanon.

---

[127] Based on a RAND colleague's discussion with a Hezbollah military analyst, Beirut, January 2009.

[128] James Dunnigan, "Hapless Hezbollah ATGMs Revealed," StrategyPage.com, September 7, 2008.

[129] David Eshel, "Hezbollah's Intelligence War: Assessment of the Second Lebanon War," Defense-Update.com, 2007.

[130] Based on a RAND colleague's discussion with a Hezbollah military analyst, Beirut, January 2009. Brun ("The Second Lebanon War," p. 9) notes that "Hezbollah's training and logistic infrastructures were located in the Lebanese Bekaa, far from the Israeli border. Syrian and Iranian supplies were received there and distributed to the various units."

Table 2.1
Hezbollah Rockets, 2006

Type	Range (km)	Payload (kg)	Quantity
Short range: 107-mm Katyusha 122-mm Katyusha 122-mm Extended Range Katyusha 240-mm Katyusha	7–40	7	13,000
Intermediate range: 240-mm Fadjr-3 330-mm Fadjr-5 220-mm Urgan 302-mm Khaibar-1	45–70	50–175	~1,000
Long range: 610-mm Zelzal 2	200	400–600	Dozens

SOURCES: Patrick Devenny, "Hezbollah's Strategic Threat to Israel," *Middle East Quarterly*, Vol. XIII, No. 1, Winter 2006, pp. 31–38; Exum, *Hizballah at War*, p. 6; GlobalSecurity.org, "Hizballah Rockets," web page, date not available; Giora Romm, "A Test of Rival Strategies: Two Ships Passing in the Night," in Shlomo Brom and Meir Elran, eds., *The Second Lebanon War: Strategic Perspectives*, Tel Aviv: Institute for National Security Studies, 2007, p. 53.

Hezbollah developed and used automatic rocket fire systems that were pre-positioned and concealed under concrete shelters prior to the 2006 war. These rockets—mainly Katyushas—were activated with an electronic trigger, and their firing angles were preset. Only a small number of fighters (in some cases, only one) had to be nearby to fire these rockets.[131] An IDF photograph of one of these well-concealed rocket systems is shown in Figure 2.2.

In addition to rocket and missile inventory listed in Table 2.1, Hezbollah had other effective standoff weapons, which are listed in Table 2.2.

Hezbollah also appears to have had fairly sophisticated capabilities beyond weaponry. For example, there are reports that it employed signals intelligence (SIGINT) "to hack into IDF communications sys-

---

[131] Author's discussions with IDF officers, Tel Aviv, March 2–5, 2008, and February 9–19, 2009.

Figure 2.2
A Concealed Hezbollah Rocket System

SOURCE: Provided to the author by the IDF in 2009.
RAND *MG1085-2.2*

tems . . . and use the information to ambush Israeli commando raids."[132]
It also intercepted radio communications between Israeli ground
commanders.[133] During the war, Israeli soldiers found Iranian-made
equipment, including "eavesdropping devices, computers and modern
communications equipment, up-to-date and detailed military maps
of Israeli strategic targets, and even lists of telephone numbers inside
Israel."[134] Furthermore, Hezbollah, anticipating Israeli electronic warfare against its networks, had used optical fibers that are not susceptible

---

[132]Nicholas Noe, "A Response to Andrew Exum's 'Hizbollah at War: A Military Assessment,'" Mideastwire.com, undated.

[133]Crooke and Perry, "How Hezbollah Defeated Israel: Part 1."

[134]Eshel, "Hezbollah's Intelligence War."

Table 2.2
Other Hezbollah Weapons

Type and Name	Maximum Range (km)
Anti-ship missiles	
C-701	15
C-802 Noor	120
Unmanned aerial vehicles	
Ababil	N/A
Antitank guided missiles	
AT-13 Metis-M	2
AT-4 Spigot	2
AT-3 Sagger	3
TOW	3.75
AT-5 Spandrel	4
AT-14 Kornet-E	5

SOURCES: Brun, "The Second Lebanon War," p. 3; Exum, *Hizballah at War*, p. 6.

to electronic jamming.[135] It also created a "significant signals-counter-intelligence capability" and was "able to predict where Israeli fighters and bombers would strike."[136] Finally, Hezbollah, "fearing Israeli eavesdropping," protected its own communications from SIGINT by "rigorously" maintaining "a high level of security and encryption. . . . Despite extensive war time activity, penetrating Hizbollah's communication systems proved a formidable challenge to Israeli intelligence."[137]

---

[135] David Eshel, "Hezbollah's Intelligence War."

[136] Crooke and Perry, "How Hezbollah Defeated Israel: Part 1."

[137] Yoaz Hendel, "Failed Tactical Intelligence in the Lebanon War."

Hezbollah used a variety of means—including UAVs, to an extent—to target Israeli military installations.[138] Hezbollah also employed human intelligence (HUMINT) to locate Israeli targets, including the IAF monitoring station at Mount Meron, which was attacked by rockets at the outset of the war. Finally, Hezbollah employed "preventive intelligence . . . against Israeli penetration into . . . [its] own network . . . [:] Hezbollah had created a special counter intelligence department tasked with enforcing organizational security."[139] Hezbollah had also "turned" a number of Israeli agents and was "able to 'feed back' false information on . . . [its] militia's most important emplacements to Israel—with the result that Israel target folders identified key emplacements that did not exist."[140]

Hezbollah's capabilities, coupled with its preparation of the battlefield and its fighters' training, presented the IDF with a formidable challenge.

How the Second Lebanon War Unfolded

The IDF responded within two hours to the July 12, 2006, abductions, sending a platoon-sized Nahal force and a Merkava tank from the 7th Brigade Armored Corps across the border in a "retaliatory attack on Hezbollah's nearest posts" and "to capture a vantage point over Giv'at Hadegel, a nearby hill on the Lebanese side, site of a Hezbollah post, and to take control of the exit roads from the village of Ita a-Sha'ab as a solution to the unlikely possibility that the abductors were still there."[141] The resources available to the 91st Division Commander were limited, and, given the amount of time that had passed since the abduction, this

---

[138] Brun, "The Second Lebanon War, 2006." Brun notes that the IAF shot down two Hezbollah UAVs (Iranian Ababils) that were on attack missions into Israel with Python 4 air-to-air missiles; a third UAV crashed inside Israel while being pursued by IAF fighters.

[139] Eshel, "Hezbollah's Intelligence War."

[140] Crooke and Perry, "How Hezbollah Defeated Israel: Part 1."

[141] Harel and Issacharoff, *34 Days*, p. 12. The Winograd Commission's detailed timeline of the war is reprinted in this book's appendix.

action was more of a retaliatory attack on nearby Hezbollah positions than a pursuit of the abductors. During this incursion, the Merkava tank was destroyed by a large IED when it attempted to run over a Hezbollah hut, and the four members of the tank's crew were killed. An additional Nahal force was sent to prevent Hezbollah from taking the bodies, but Hezbollah attacked this force with mortar fire, killing another IDF soldier.[142]

At the time of the abduction, Hezbollah also attacked Israeli military outposts and the towns of Shlomi and Shebaa Farms with rockets, an action that is now viewed as an attempt to divert Israeli attention during the kidnapping.[143] Hamas and other groups were also launching rockets at Israel.[144] Finally, the IDF was also involved at the time in large-scale operations in Gaza to attempt to recover kidnapped Israeli soldier Gilad Shalit. Indeed, the Hezbollah and Shalit abductions were viewed by Israeli leaders as part of a pattern of terrorist activity that would only get worse if it was not dealt with aggressively.[145]

On July 12, the IDF further responded to the Hezbollah abductions with preplanned air strikes against 17 Hezbollah locations in Lebanon and against three bridges that spanned the Litani River. The attacks on bridges were to prevent the movement of the abducted soldiers out of southern Lebanon.[146]

Israeli political and military leaders met to discuss what to do about the abduction and rocket attacks. They agreed that Israel needed

---

[142] Harel and Issacharoff, *34 Days*, pp. 12–13. Harel and Issacharoff note that "'Nahal' is a Hebrew acronym for Pioneer Fighting Youth" (p. 264). See also William M. Arkin, *Divining Victory: Airpower in the 2006 Israel-Hezbollah War*, Maxwell Air Force Base, Ala.: Air University Press, 2007, pp. 1–2.

[143] Kenneth R. Bazinet and Helen Kennedy, "'An Act of War': Israel Reacts to Slay of 8, Sends Tanks to Lebanon," *New York Daily News*, July 13, 2006, p. 6.

[144] For example, see "Palestinian Groups Claim Rocket Attacks Against Israeli Targets," *BBC Monitoring Middle East*, June 22, 2006; and "Islamic Jihad Armed Wing Rejects Abbas Calls to Halt Rocket Fire," *BBC Monitoring Middle East*, June 23, 2006.

[145] Author's conversation with an IDF officer via telephone, January 17, 2011.

[146] Arkin, *Divining Victory*, pp. 3–4.

to respond with military force.[147] However, both the political and the military leadership expressed reluctance to send a large ground force into Lebanon, preferring instead to rely on airstrikes and limited ground raids on towns of symbolic importance. This decision meant that the on-the-shelf plan for a war in southern Lebanon—Operation Sky Water (in Hebrew, Mei Marom)—was not executed. This plan called for a massive Israeli Army sweep to "isolate the southern part of Lebanon from northern sources of supply, and then [to] eradicate the Hezbollah in the sealed off area."[148] An alternative plan—Operation Ice Breaker (in Hebrew, Shoveret Kerach)—"was based on standoff fire and limited cross-border raids," was also not fully employed.[149]

Indeed, as late as July 18, General Halutz, the IDF chief of staff, stated, "There is no point planning a ground operation . . . the method is not one of divisional operations. It is necessary to take the direction of smaller, shorter, and more focused things."[150] Initially, Israeli leaders also decided against calling up reserves.

Israel's Strategy in Lebanon

At the beginning of the conflict in Lebanon, the IDF had very limited goals. General Halutz testified before the Winograd Commission that, "during the initial stages of the operation[,] he was thinking in terms of a retaliatory attack, not war. He therefore instructed his subordinates at the General Staff level to refrain from relating to the operation as war."[151]

A July 12 communiqué, issued after an emergency meeting of the Israeli Security Cabinet, framed the initial Israeli response to the soldiers' abductions:

---

[147] Cordesman, Sullivan, and Sullivan, *Lessons of the 2006 Israeli-Hezbollah War*, p. 6; Harel and Issacharoff, *34 Days*, p. 85.

[148] Gil Merom, "The Second Lebanon War: Democratic Lessons Imperfectly Applied," *Democracy and Security*, Vol. 4, No. 1, January 2008, p. 17.

[149] Merom, "The Second Lebanon War," p. 17.

[150] The Winograd Commission, p. 76.

[151] Avi Kober, "The Israel Defense Forces in the Second Lebanon War," p. 9.

Israel views the sovereign Lebanese Government as responsible for the action that originated on its soil and for the return of the abducted soldiers to Israel. Israel demands that the Lebanese Government implement UN Security Council Resolution #1559.

Israel will respond aggressively and harshly to those who carried out, and are responsible for, today's action, and will work to foil actions and efforts directed against it. . . . Israel must respond with the necessary severity to this act of aggression and it will indeed do so.[152]

Some Israelis believe that the lack of significant military experience among the senior political figures in the Israeli Security Cabinet hampered development of a coherent strategy.[153] However, the three major players in the cabinet did have some military experience. Prime Minister Ehud Olmert had served in the infantry, Defense Minister Amir Peretz had reached the rank of captain in the paratroopers, and Foreign Minister Tzipi Livni had served as a lieutenant and instructor at a training institution. Still, compared with past Israeli political leaders who were career military officers (such as former IDF Chiefs of Staff Shaul Mofaz and Ehud Barak), these three had little experience working with senior-level military officers or developing broad military strategy prior to their political careers.[154] Nevertheless, as the Winograd Commission notes, the military, and particularly General Halutz,

---

[152] Israeli Security Cabinet communiqué of July 12, 2006, quoted in Arkin, *Divining Victory*, p. 6. Much of the key decisionmaking in the war was made by a small group of cabinet members known as the "Group of Seven."

[153] For example, see Harel and Issacharoff, *34 Days*, pp. 136–137. Harel and Issacharoff write that both Prime Minister Ehud Olmert and Defense Minister Amir Peretz "often repeated the excuse, 'I am not a general.'"

[154] For biographical information on Omert, Peretz, and Livni, see Ami Isseroff, "Biography of Ehud Olmert," Zionism-Israel.com, updated September 2008; Yaakov Katz, "Defense Officials Debate Peretz's Rumored Appointment," *The Jerusalem Post*, April 5, 2006, p. 2; The State of Israel, "Ehud Olmert," web page, 2009; Leslie Susser, "Marching to the Livni Beat," *The Jerusalem Post*, July 7, 2008, p. 14; Elli Wohlgelertner, "The Mayor's Grand Plan," *The Jerusalem Post*, December 5, 1997, p. 14; and Phil Zabriskie, "The Former Dove Who's Directing Israel's War," *Time*, July 17, 2006.

did little to compensate for the lack of expertise among Israeli politicians. This became apparent as the war unfolded:

> The COS [IDF Chief of Staff] did not alert the political echelon to the serious shortcomings in the preparedness and the fitness of the armed forces for an extensive ground operation, if that became necessary. In addition, he did not clarify that the military assessments and analyses of the arena . . . [indicated] that there was a high probability that a military strike against Hezbollah would make such a move [ground operations] necessary.[155]

One Israeli officer noted that the military objectives were mainly focused on deterrence: "In essence, Israel's government decided to conduct a military operation in order to convey a clear message that would prevent future kidnappings, rather than to wage war."[156] Giora Eiland, former head of Israel's National Security Council, made a less-charitable assessment: "Intelligence told how bad the Arabs are. The ministers asked the army what it intended to do. The officers answered: We'll attack and see what happens. And that's how it was."[157]

A July 12 order framed how the IDF would prosecute operations in Lebanon in the coming weeks:

> The operational order issued on July 12 reflects the military's understanding of the specific list of missions assigned to them. This order instructed the IDF to destroy Hezbollah's long-range rocket launchers and to damage the organization's launch capability, attack its soldiers, attack its commands and infrastructure, strike its symbols and assets, and destroy Hezbollah infrastructures next to the Israeli border in order to establish a special secu-

---

[155] "Excerpts from the Winograd Report," YNetNews.com, April 30, 2007.

[156] Brun, "The Second Lebanon War," p. 12. See also Efraim Inbar, "How Israel Bungled the Second Lebanon War," *Middle East Quarterly*, Summer 2007. Inbar notes that

> unrealistic goals compounded poor preparation. Israeli political and military leaders erred in their belief that Israeli pressure on Hezbollah and the weak Lebanese government could generate a political process in which the Lebanese army could achieve a monopoly over the use of force in Lebanon.

[157] Harel and Issacharoff, *34 Days*, p. 90.

rity zone. At the same time, the IDF was instructed to set up an aerial and naval blockade to prevent Syria and Iran from supplying arms to Hezbollah.[158]

On July 19, the specifics of Israel's strategy to reach its desired political end state were determined by the Ministerial Committee on National Security Affairs:

> The principles for the political solution of the crisis, in the Lebanese theater will be:
>
> 1) Release of the abducted soldiers and returning them to Israel unconditionally;
>
> 2) Stopping of the firing of missiles and rockets against the residents of the State of Israel and against Israeli targets;
>
> 3) Full, complete application of Resolution 1559 of the United Nations, including disarming all of the armed militias, applying the sovereignty of the Government of Lebanon to its entire territory and deployment of the Lebanese Army along the border with Israel.[159]

This strategy required the Lebanese government to do something it clearly lacked the ability to do, regardless of its inclinations. Furthermore, the ability of the IDF to employ military force to achieve these political objectives would soon be tested.

The Air War

During meetings on July 12, the Israeli government decided to execute a bombing campaign to knock out Hezbollah's Fajr intermediate-range rockets (Operation Specific Weight), shut down the runways of the Beirut Rafic Hariri International Airport, interdict the Beirut-

---

[158] Brun, "The Second Lebanon War," p. 14. Brun also notes that "a ground offensive was not discussed seriously on the first day. When the topic was raised in the following days, the majority of decision makers, both political and military, opposed it."

[159] The Winograd Commission, p. 82.

Damascus highway, and attack the al-Manar television station.[160] General Halutz, who had advocated an even broader attack that included the Lebanese electric grid, believed that the IDF had to respond vigorously and rapidly because "Israel had a limited margin of time, two to three days at most, before the international community intervened."[161]

Clearly, air strikes were central to a plan that its designers assumed would damage Hezbollah, pressure the Lebanese government to control Hezbollah, result in the release of the captured soldiers, and strengthen Israel's military deterrent.[162] Furthermore, there was little appetite among civilian ministers for a large-scale ground campaign, as Defense Minister Peretz recalled after the war:

> No one was eager to bring in, to confrontation zones, land forces. Especially . . . not as the Lebanese trauma is hovering over everybody's head. The truth needs to be said. The trauma of exiting Lebanon was hovering over the heads of the Cabinet ministers and some of the leaders of the Army. It delayed the decision on land intervention. . . . Most [ministers] reiterated once and again their adamant opposition to a large scale land intervention. Instead, there were calls to erase villages rather then get [land] forces in; to turn villages into soccer fields and sand.[163]

Prime Minister Olmert also seemed to assume that the decisive air operation would force Hezbollah to sue for a ceasefire after only a

---

[160] Harel and Issacharoff, *34 Days*, p. 86. The airport and highway bombings were ostensibly planned to prevent the movement of the abducted soldiers out of the country. See also David Makovsky and Jeffrey White, *Lessons and Implications of the Israel-Hizballah War: A Preliminary Assessment*, Policy Focus No. 60, Washington, D.C.: The Washington Institute for Near East Policy, 2006, p. 19; and The Winograd Commission, p. 585. The longer-range Zelzals were hit on July 18.

[161] Harel and Issacharoff, *34 Days*, p. 78.

[162] Brun, "The Second Lebanon War," p. 13.

[163] *Haaretz* interview with Ari Shavit, May 3, 2007, cited in Merom, "The Second Lebanon War," p. 22. The "Lebanese trauma" noted in the interview refers to Israel's experiences in the First Lebanon War and the to the prolonged Israeli occupation of southern Lebanon, which lasted until 2000.

few days, "during which time Israel would weather Katyusha fire."[164] General Halutz was likely responsible for the prime minister's optimism, having stated his belief that "the battering would last two or three days."[165]

The initial strikes were executed, with great success, on July 13. The IAF also blockaded Lebanese airspace, and the Israeli Navy blockaded Lebanese ports. Hezbollah responded to these attacks with an intense barrage of Katyusha rockets aimed at Israeli border settlements,[166] and, for the first time, it fired rockets at Haifa, an Israeli city 30 km from the Lebanese border.[167]

On July 14, senior Israeli cabinet members met and approved an escalation in the form of IDF plans to bomb Dahiye, the southern Beirut suburb that housed Hezbollah's headquarters and Secretary-General Nasrallah's private residence.[168] Proponents of the attack believed it would "deliver a painful and surgical blow to Hezbollah's strength and honor."[169] A minority at the meeting disagreed, and Israeli journalists Amos Harel and Avi Issacharoff believe that "some participants realized that an attack on Dahia presaged all-out war on Hezbollah. The idea of 'bang and it's over' increasingly receded."[170] Events would soon prove the skeptics correct.

---

[164] Harel and Issacharoff, *34 Days*, p. 84 (see also p. 81). When Prime Minister Olmert talked to U.S. Secretary of State Condoleezza Rice on July 12, she "gave a green light" to the Israeli strike plans. Secretary Rice also "brought up two American non-gos regarding the Israeli operation: Siniora must not be hurt and the civilian infrastructure must not be bombed." Prime Minister Olmert agreed to these demands. The United States viewed the Siniora government as key to future Lebanese stability and did not want to see it toppled as the result of Israeli military action.

[165] Harel and Issacharoff, *34 Days*, p. 87.

[166] Harel and Issacharoff, *34 Days*, p. 87.

[167] BBC News, "Rockets Hit Israeli City of Haifa," News.BBC.co.uk, last updated July 13, 2006.

[168] Harel and Issacharoff, *34 Days*, pp. 99–100.

[169] Harel and Issacharoff, *34 Days*, p. 100.

[170] Harel and Issacharoff, *34 Days*, p. 100.

The IAF began striking Dahiye on the evening of July 14, and the suburb was "transformed into a sea of rubble."[171] In the coming weeks, the IAF stepped up its air attacks, striking Hezbollah command posts, bridges, traffic intersections, ammunition caches, and rocket launchers in southern Lebanon. Convoys suspected of carrying munitions between Syria and Lebanon were also struck. Artillery and naval bombardments were used to attack targets in Lebanon.[172] Air attacks continued until the conclusion of the war on August 14. The scale of air effort was enormous:

> During the war the IAF's fighters and helicopters carried out about 19,000 sorties, of which some 12,000 were fighter sorties in attack and support missions (an average of about 300 sorties a day). Attack and transport helicopters carried out some 4,200 additional sorties. Most of the air activity was directly aimed at Hezbollah. During the war, the IAF attacked about 7,000 targets, using about 19,000 bombs and about 2,000 missiles; approximately 35 percent of the ammunition used during the war was precision-guided munitions (PGMs). In parallel to the kinetic operations, the IAF also operated in the information realm and dropped about 17.3 million flyers.[173]

Figure 2.3 shows the locations struck by the Israelis during the war. Indeed, the Winograd Commission notes that "the scope of the Air Force's operations was unprecedented and absolutely disproportionate to Israel's past wars. All in all, there were more assault sorties and more flight hours than during the Yom Kippur War."[174] Additionally, the Israeli Army fired 150,000 rounds (of 155-mm rounds and rockets), and the Israeli Navy fired more than 2,500 rounds.[175] By July 16, the

---

[171] Harel and Issacharoff, *34 Days*, p. 101.

[172] Brun, "The Second Lebanon War," p. 16; Catignani, *Israeli Counter-Insurgency and the Intifadas*, p. 189.

[173] Brun, "The Second Lebanon War," p. 2.

[174] The Winograd Commission, p. 314.

[175] Catignani, *Israeli Counter-Insurgency and the Intifadas*, p. 189.

Figure 2.3
Map of Lebanese Targets Attacked by Israel

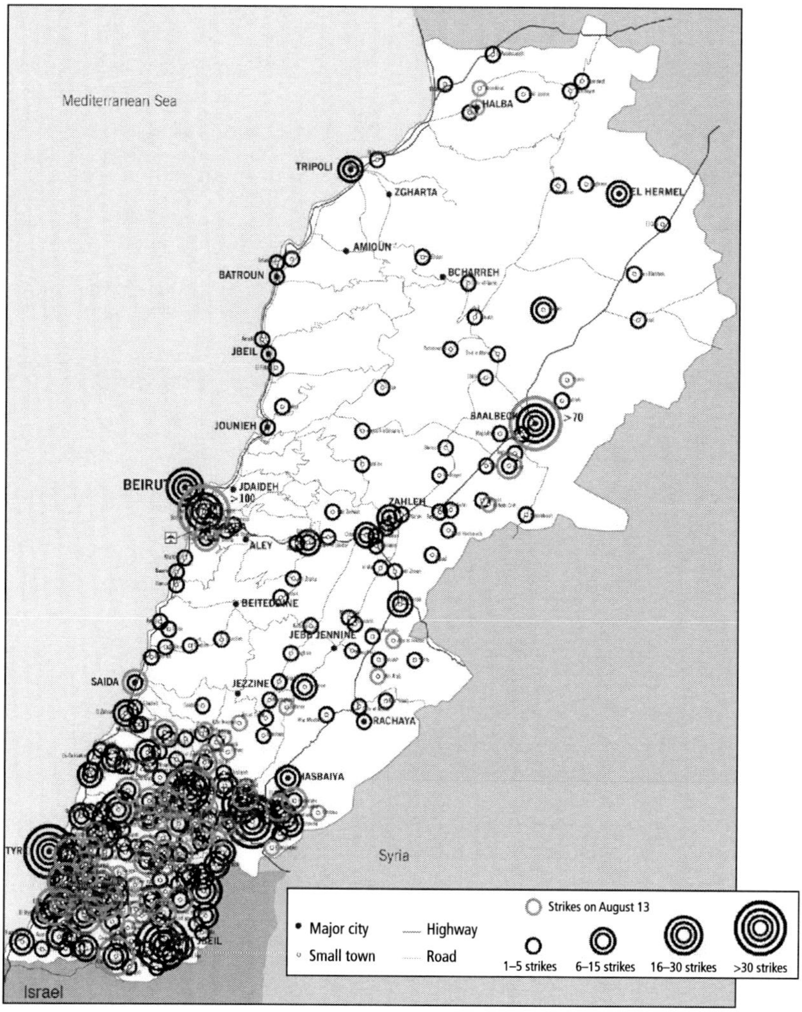

SOURCE: Samidoun Media Team.

RAND MG1085-2.3

IDF literally ran out of targets, having "battered all 83 objectives" on Northern Command's initial target list.[176]

Although much has been written about the Israeli "overreliance" on air power during the Second Lebanon War, author William Arkin lays out a different view:

> Israeli politicians had many valid reasons to want to use the air-power tools associated with strategic attack and long-range strike. First, an "airpower"-centric approach best countered the enemy's strengths, particularly given how embedded Hezbollah was in Lebanese civil society and how much it had built up its basic capabilities north of the Litani River (and thus out of the reach of Israeli ground forces). Second, the existing conception of conventional ground combat, attrition, and occupation prevalent in the IDF was out of synch with either the nature of the enemy or the level of commitment Israeli leaders (and, in their view, the Israeli public) were willing to make. Third, the "airpower" decision was made easier if not de facto by the stark reality that the ground forces were not prepared to mount the very campaign they were promoting.[177]

The IAF enjoyed much more success in targeting Hezbollah's intermediate- and long-range rockets than in targeting its short-range arsenal.[178] The IAF had intelligence, provided by the General Security Service, the Mossad, and Israeli Military Intelligence, on the long-range launchers, and it also had existing plans for attacking them.[179]

During the war, the IAF adapted its techniques in order to find and destroy intermediate- and long-range launchers. Adaptations

---

[176] Author's discussions with IDF officers, Tel Aviv, March 2–5, 2008; Harel and Issacharoff, *34 Days*, p. 122.

[177] Arkin, *Divining Victory*, p. 153. For a contrasting view, see Matt M. Matthews, *We Were Caught Unprepared*; Matthews, "Hard Lessons Learned."

[178] Cordesman, Sullivan, and Sullivan, *Lessons of the 2006 Israeli-Hezbollah War*, p. 10. See Makovsky and White, *Lessons and Implications of the Israel-Hizballah War*, p. 40, for a detailed breakdown of the rockets and antitank weapons reportedly in the possession of Hezbollah.

[179] Harel and Issacharoff, *34 Days*, p. 91; The Winograd Commission, p. 314.

included using overhead sensors, special forces, and UAVs to both locate and track launchers and their crews and to carry out air strikes designed to destroy them.[180] Persistent ISR, particularly from UAVs, was very important in this effort. Nevertheless, throughout the conflict, Hezbollah was able to launch rockets into Israeli territory, launching an average of 90–150 attacks a day.[181] Most of these attacks involved short-range Katyushas that the IAF was not able to find with air ISR platforms. The continual rain of rockets into Israel shocked the Israeli population and undermined the image of the IDF as an effective military force.

Quite simply, as the Winograd Commission determined, "the Air Force was unable to strike the short-range Katyusha alignment in a way that would limit the attacks on the Israeli home front." The commission also carefully points out that "the Air Force had reached this assessment of its own capability in this area even before the war."[182] Thus, the issue confronting Israeli decisionmakers was how to stop the short-range rockets. The minutes of a meeting of the General Staff on August 4 show that there was a growing recognition that standoff fires would not stop the short-range rockets. The head of the Intelligence Branch, with support from the head of the Research Division, stated, "I thought that two weeks of work by the Air Force would lead us to another place, and today I think differently. I think that the only way of clearing these 'Katyusha rockets' . . . is with the help of a significant ground operation."[183] As will be seen, this was the course that was eventually adopted, with very mixed results.[184]

---

[180] Author's discussions with IDF officers, Tel Aviv, March 2–5, 2008, and February 9–19, 2009.

[181] Brun, "The Second Lebanon War," 2006, pp. 16, 19, 21. Brun divides the war into three phases: Phase I, July 12–19; Phase II, July 20–31; and Phase III, August 1–14. The average number of Hezbollah rocket attacks on Israel were, by phase, 90 (Phase I), 115 (Phase II), and 150 (Phase III).

[182] The Winograd Commission, p. 315.

[183] The Winograd Commission, p. 153.

[184] Hezbollah's ability to continue firing short-range rockets is reminiscent of the German V-1 and V-2 rocket attacks against the United Kingdom and elsewhere (particularly Antwerp, Belgium) during World War II. Although Allied bombing raids destroyed some of the

The scale of the Israeli air attack surprised observers both inside and outside the Middle East. The international community denounced Israel for attacking Lebanon's infrastructure. Secretary-General Nasrallah was surprised by the "size and strength of the response."[185] Despite the air attack, however, Hezbollah was able to surprise the IDF in return. For example, on July 14, Hezbollah hit the *Hanit*, an Israeli Sa'ar 5–class corvette—one of Israel's most advanced ships—with an Iranian-made C-802 Noor missile.[186] The Israeli chief of naval operations said that the Israelis "were not aware that Hezbollah possessed this kind of missile."[187] The ship's crew, not expecting to be attacked, had turned off its Barak antimissile system.[188]

The Initial Ground War

It is clear that Hezbollah was well prepared for an Israeli incursion into Lebanon. Initial Israeli ground actions following the July 12 abductions were mainly limited raids. By July 14, the Shaldag, Egoz, and Yamam units (consisting of elite special operations forces) had taken control of part of the village of Rajar. On July 17 and 18, there were more Israeli raids 1–2 km inside Lebanon. These were aimed at destroying Hezbollah positions along the border.[189]

It is also clear that the IDF had little understanding about the opponent it was facing. Unlike the IAF, which had intelligence useful in targeting Hezbollah's long-range rockets, the Israeli Army went into

---

German capacity to launch rockets, the Allied invasion of Europe and their subsequent over-running of the rocket sites (or the fact that the Germans moved the sites out of harm's way and, hence, out of range of the United Kingdom and other key targets) are what stopped this threat.

[185] Arkin, *Divining Victory*; "Nasrallah Admits 'Intelligence Error,'" *The Jerusalem Post*, February 3, 2007.

[186] Harel and Issacharoff, *34 Days*, p. 101.

[187] James Gordon, "Iran Called Source of Missile that Struck Ship," *New York Daily News*, July 19, 2006, p. 19.

[188] Yaakov Katz and Sam Ser, "IDF Report Card," *The Jerusalem Post*, August 25, 2006, p. 16.

[189] Harel and Issacharoff, *34 Days*, pp. 127, 371.

Lebanon largely without current intelligence. A postwar assessment notes that,

> upon mobilization of ground forces in Lebanon and the addition of other units, a shortage of current information emerged. Moreover, where such information existed, it did not reach those who needed it, remaining untapped. For example, the sealed boxes prepared in advance by [Military] Intelligence that included relevant information for ground forces remained unused, while the "obsolescence" of aerial photos, dating back to 2002, featured prominently on the list of complaints raised by reservists. Information that was available did not reach its target due to inefficient information pipelines.[190]

Combat commanders' lack of accurate information was particularly problematic in the case of the so-called nature reserves. Colonel Haggai Moredechai, commander of the Paratroopers Brigade, stated that, until the IDF encountered the extensive Hezbollah positions, "We thought we were dealing with [people with] a few pup tents, with sleeping bags and cans of food."[191]

The first major ground engagement between the IDF and Hezbollah took place in Maroun al-Ras, a town near the Israeli border. On July 17, soldiers from the IDF's elite Maglan unit began probing near Maroun al-Ras and soon encountered a Hezbollah ambush. The Israeli

---

[190] Hendel, "Failed Tactical Intelligence in the Lebanon War." Hendel also notes that part of the reason for Israel's poor tactical intelligence in Lebanon was the fact that the tactical intelligence apparatus was focused on dealing with the intifada on the "Palestine front."

[191] Jewish Daily Report, "Israel Prepares for War with Lebanon/Hizbollah Terrorists," JewishDailyReport.wordpress.com, July 23, 2010. The article also notes that

> the IDF knew about the nature reserves in Lebanon from which Hezbollah launched rockets in 2006 before the last war, but much of the relevant information did not make its way to the combat units. On July 19, 2006, when the elite unit Maglan found itself in the heart of the first Lebanese nature reserve the IDF encountered in the war—on a hill overlooking Moshav Avivim, just across the border—the detailed intelligence about the site sat in sealed crates: Military Intelligence had not made it available to the forces in time, claiming it was too highly classified.

soldiers were surprised by Hezbollah's preparations and its forces' fighting skills:[192]

> "We didn't know what hit us," said one of the soldiers, who asked to be named only as Gad. "In seconds we had two dead." With several others wounded and retreating under heavy fire the Maglans, one of the finest units in the Israeli Defence Forces (IDF), were astonished by the firepower and perseverance of Hezbollah.
>
> "Evidently they had never heard that an Arab soldier is supposed to run away after a short engagement with the Israelis," said Gad.
>
> "We expected a tent and three Kalashnikovs—that was the intelligence we were given. Instead, we found a hydraulic steel door leading to a well-equipped network of tunnels."[193]

The IDF began piling on in Maroun al-Ras, sending tanks from multiple units, the "Egoz" reconnaissance unit from the Golani Brigade, an engineering battalion, and a battalion of the Paratroopers Brigade. The IDF lost five soldiers in the fighting in and around Maroun al-Ras, and the town was not secured until July 23.[194]

On July 23, IDF forces advanced further into southern Lebanon, engaging Hezbollah militants near Bint Jbeil, the largest town near the Lebanese-Israeli border. IDF units engaged in the action included the 51st Battalion of the Golani Brigade, the 101st Battalion of the Paratroopers Brigade, and forces from the armored corps' 7th Brigade.[195]

---

[192] Crooke and Perry, "How Hezbollah Defeated Israel: Part 1"; Uzi Mahnaimi, "Humbling of the Supertroops Shatters Israeli Army Morale," TimesOnline.co.uk, August 27, 2006; Anshel Pfeffer, "After Maroun al-Ras Battle, Bint-Jbail Looms as Next Challenge," *The Jerusalem Post*, July 24, 2009.

[193] Mahnaimi, "Humbling of the Supertroops."

[194] Matthews, "We Were Caught Unprepared," pp. 43–44; Ned Warwick, "Battle-Tested: Young Israelis, Fresh from Front, Recount Clash," *The Philadelphia Inquirer*, July 25, 2006, p. A1. See also Exum, *Hizballah at War*, p. 9; Israel Ministry of Foreign Affairs, "The IDF's Secret Weapon Against Hizbullah," article, December 5, 1996.

[195] See "Hezbollah Leader Calls for Muslim-Christian Coexistence," Lebanese Broadcasting Corporation, May 26, 2000, in *BBC Summary of World Broadcasts*, May 27, 2000. Bint

At first, the IDF was only going to conduct a raid, attacking from the north and killing as many Hezbollah fighters as it could before leaving town. Instead, General Halutz ordered that the troops "occupy" the town, and plans were changed to have the attack come from the south. The Israeli attackers ran into heavy resistance and lost eight soldiers. Fighting in the town continued for several days before the IDF withdrew.[196] As the Winograd Commission later acknowledged, "The fighting in Bint Jbeil did not yield the desired achievement and became a symbol of the unsuccessful action of the Israel Defense Forces throughout the fighting."[197]

At this point in the conflict, four Israeli divisions had entered the fighting in Lebanon: the 91st Division (Infantry, "Galilee Formation"), the 162nd Division (Armored, "Steel Formation"), the 366th Division (Reserve Armored, "Pillar of Fire Formation"), and the 98th Division (Paratroopers, "Fire Formation"). A brief summary of their actions through August 10 is provided here:

- 91st Division. The division was in charge of the entire sector until July 21, when it transferred the eastern sector to the 162nd Division. Its main area of fighting was in and around Bint Jbeil.
- 162nd Division. The division fought in the eastern sector beginning on July 21. Its two-brigade attack on al-Khiam ended on July 25.
- The 98th Division. The division began operations in the central sector on July 30.
- The 366th Division. The division took over responsibility for the eastern sector of the 162nd Division's sector on August 4. It fought in the area of Marj Ayoun and al-Khiam between August 8 and August 10.[198]

Jbeil, a town of some 30,000 residents, became famous in 2000 when Hezbollah Secretary-General Nasrallah spoke from the town on the occasion of the Israeli withdrawal from Lebanon. It was there that he declared that Israel was "weaker than a spider web."

[196] Harel and Issacharoff, *34 Days*, pp. 138–143.

[197] The Winograd Commission, p. 305.

[198] The Winograd Commission, p. 305.

All four divisions primarily conducted battalion- and brigade-sized raids against Hezbollah positions and rocket locations. There was not a coordinated Northern Command campaign to accomplish a collective objective linked to a strategic end state.

Losing Support for the War

Several incidents combined to put external and internal pressure on Israel to end operations in Lebanon. The first occurred on July 30, when an IAF airstrike that destroyed a building in Qana, a town located 10 km south of Tyre, killed more than 20 people, many of them children. This incident drew both the ire of the international community and charges that the IDF was deliberately targeting civilians.[199] Qana was a turning point in international opinion. Prior to the airstrike, there had been some level of support for Israel among the international community, and the leaders of the Group of Eight had issued a condemnation of Hezbollah's actions on July 16.[200] In the aftermath of the incident, the United States began pressing Israel to reach a ceasefire.[201] On August 5, the United States and France introduced the text that would ultimately become UN Resolution 1701, which was approved unanimously by the UN Security Council on August 11.[202]

After Qana, there was a growing sense among Israelis that they were not going to win in Lebanon.[203] It was two other events, however, that turned Israeli domestic opinion against the war. On August 6, a Katyusha rocket launched by Hezbollah killed 12 IDF paratroopers at Kfar Giladi. The incident was broadcast on Israeli media and "had an immense impact on public morale."[204] Later that same day, a Hez-

---

[199]Colum Lynch, "U.N. Voices 'Extreme Shock' over Israeli Strike," *The Washington Post*, July 30, 20006, p. A9.

[200] The Winograd Commission, p. 584.

[201] Harel and Issacharoff, *34 Days*, pp. 165–170.

[202]United Nations Security Council, "Security Council Calls for End to Hostilities Between Hizbollah, Israel, Unanimously Adopting Resolution 1701," press release SC/8088, August 11, 2006.

[203]Author's conversation with an IDF officer via telephone, January 17, 2011.

[204]Harel and Issacharoff, *34 Days*, p. 188.

bollah rocket hit Haifa, killing three civilians. One of Prime Minister Olmert's staff members later recalled that

> this was the moment when public opinion switched. . . . Two days earlier, there was a feeling that the whole country was united. But on August 6, a sense of distaste and disgust sunk in. People had the impression that we were "catching it" on every corner, almost like the destruction of the Third Temple.[205]

Time was running out for Israel to reach a satisfactory conclusion to the conflict.

One Final Offensive

Planning for a major ground operation in southern Lebanon began on August 3 at the direction of Defense Minister Peretz. The plan (Operation Changing Direction 11) was approved by Defense Minister Peretz on August 7 and by the Israeli Security Cabinet on August 9.[206] Indeed, several preparatory moves for the envisioned operation (Operation Changing Direction 8) were begun overnight on August 8–9.[207] Changing Direction 11 was originally scheduled to begin on August 9 but was delayed until two days later. The goals of the operation were essentially two-fold. The first was to create conditions on the ground that would result in a UN Security Council resolution on the war favorable to Israel by having IDF forces push toward the Litani River.[208] The

---

[205]Harel and Issacharoff, *34 Days*, pp. 188–189.

[206]See Yaakov Katz, "Security and Defense: The Story of 'Changing Direction 11,'" JPost. com, October 1, 2008. Katz notes that "the '11' corresponds to the number of updates that were made to the plan for that offensive."

[207]The Winograd Commission, p. 366.

[208]See also Gidi Waits and Uri Blau, "General Mofaz Runs for Office," *Haaretz*, August 21, 2008. At the Israeli Security Cabinet meeting where Operation Changing Direction 11 was approved, Transportation Minister Shaul Mofaz (former IDF Chief of Staff and defense minister) offered an alternative plan: "You want to reach the Litani [River]? Then conquer the Litani. You will reach it within 48 hours and say that we won and that southern Lebanon is encircled. If you want, mop up the area from south to north." At this meeting, the cabinet also authorized Prime Minister Olmert and Defense Minister Peretz to decide the timing of the operation (The Winograd Commission, p. 594).

second was to reduce the volume of rocket fire on Israel.[209] In short, as the Winograd Commission notes, Operation Changing Direction 11 "was meant to be a large, broad ground operation, which would fundamentally alter reality in Southern Lebanon and the image of the operation in the military sense."[210] Defense Minister Peretz pressured Prime Minister Olmert to approve the operations to ensure that the war would not end "with the image of Israel as the losing side."[211]

As discussed earlier, ground operations until this point generally had been executed by brigades and smaller units and are best characterized as raids. Israeli special operations forces were also involved, having conducted a number of raids into Lebanon. One such raid was an August 5 naval commando unit action in Tyre, during which 27 Hezbollah and Iranian Revolutionary Guard Corps operatives were killed. At the time of this raid, Israeli special operations forces had conducted at least 17 operations inside Lebanon, including an August 2 action undertaken by Sayeret Matkal and Shaldag commandos on a Hezbollah headquarters facility in Baalbeck during which ten Hezbollah members were killed and five were captured.[212]

Operation Changing Direction 11 began at 1700h (Tel Aviv time) on August 11, even as final negotiations on the UN-brokered ceasefire were nearing a conclusion. It was the first large-scale IDF coordinated ground action of the war. Perhaps the most problematic aspect of Operation Changing Direction 11 was the fact that, given the impending ceasefire, Prime Minister Olmert allotted only 60 hours for the operation.[213] The original plan, presented by Major General Gadi

---

[209] Katz, "Security and Defense." See also The Winograd Commission, p. 175. General Halutz was very specific regarding rockets: "This operation [Changing Direction 11] is an operation whose purpose is to reduce the launching of short-range rockets. Not cancel, not terminate, not prevent, not crush, not shatter or anything. Reduce, how much is reducing? Less than now."

[210] The Winograd Commission, p. 370.

[211] Harel and Issacharoff, *34 Days*, p. 212.

[212] Efrat Weiss, "Tyre Raid 'Heroic Operation,'" YNetNews.com, August 5, 2006.

[213] See Harel and Issacharoff, *34 Days*, p. 213. Prime Minister Olmert explained both the 60-hour time limit and why such a large-scale operation was planned so near the end of the

Eizenkot, then head of IDF Operations Directorate, had envisioned a much longer campaign:

> The first stage—during which the forces will reach their positions—will take four days; the second stage of fighting and bringing results will take three to four weeks; and the third stage is the week it will take to withdraw the troops.[214]

Operation Changing Direction 11 was a four-division operation. In broad outlines, the plan envisioned the 91st Division pressing north, the 162nd Division crossing the Saluki River from the east and pushing north (where it would link up with the 98th Division along the Litani River), the 98th Division conducting a helicopter assault and moving to the Litani River in the direction of Tyre, and the 366th Division attacking Hezbollah forces north of Metula.[215] The scheme of maneuver for Operation Changing Direction 11 is shown in Figure 2.4. The divisions' objectives and outcomes before the cessation of hostilities at 0800h (Tel Aviv time) on August 14 are detailed in the subsections that follow.

91st Division

The 91st Division was supposed to "sweep through the eastern sector of southern Lebanon,"[216] taking and holding key terrain to prevent rocket launches and conducting "clearing and sweeps."[217] Its principal role was "to serve as a backstop for the primary effort attack" made by the 162nd Division, which would come from the north.[218] Part of the 91st Division's mission was to complete the occupation of Bint Jbeil and Eita a-Sha'ab—an objective that it did not accomplish. The over-

---

war thus: "On Monday at 7:00 [a.m.], the cease-fire is supposed to go into effect, and the UN document is unfavorable to us. We hope to better it."

[214] Katz, "Security and Defense."

[215] Harel and Issacharoff, *34 Days*, p. 194; The Winograd Commission, p. 369.

[216] Katz, "Security and Defense."

[217] The Winograd Commission, p. 369.

[218] The Winograd Commission, p. 304.

Figure 2.4
Ground Scheme of Maneuver: Operation Changing Direction

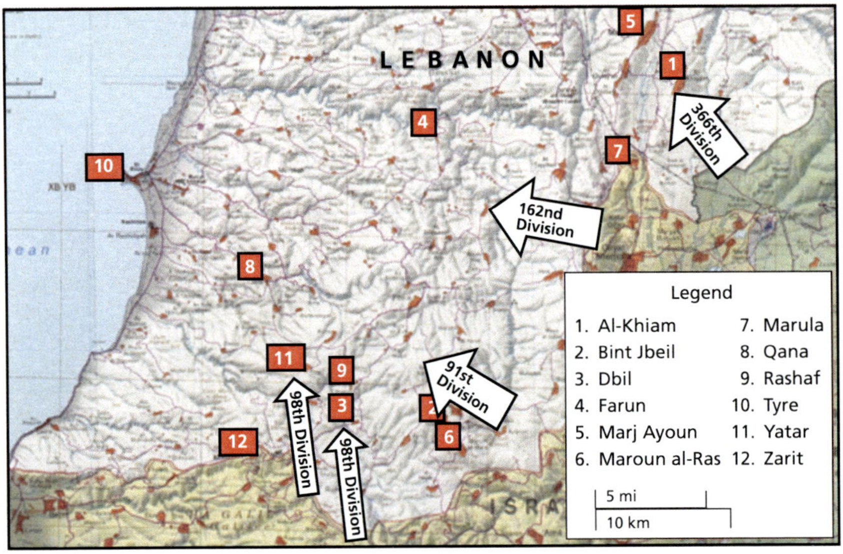

SOURCE: Adapted from "Southern Lebanon Border Area (1986)," courtesy of the
University of Texas Libraries, The University of Texas at Austin.
RAND *MG1085-2.4*

all performance of the division had little effect on the outcome of the
operation, however, the 98th Division failed to accomplish its mission.

162nd Division

The 162nd Division constituted Northern Command's main effort in
Operation Changing Direction 11. Its mission was to take up positions
throughout the western sector of southern Lebanon.

The division began operations to breach the Wadi Saluki on
August 9 but stopped these activities when Operation Changing
Direction 11 was delayed. On August 11, the division began clearing
missions along the border. On August 12, the division's 401st Armored
Brigade began offensive operations to cross Wadi Saluki. It was a risky
operation:

> Crossing the Saluki required the troops and tanks [to] climb a
> steep hill overlooked by mountains in every direction. . . . Hiz-

bullah knew that the only passage West was through the Saluki. At least 100 guerrillas took up positions with the most advanced anti-tank missile—the Russian-made Cornet—and waited.[219]

The division commander, aware of the risks, had positioned infantry soldiers from the Nahal Brigade near the villages of Andouriya and Farun to cover the armored brigade's passage. A column of tanks from the 401st Armored Brigade's 9th Battalion approached a crossing and was ambushed by a well-armed and well-prepared Hezbollah force. The 9th Battalion's forces were supposed to be supported by members of the 931st Nahal Battalion, but coordination between the two units was poor, even though their respective brigade commanders were stationed in the same house.[220] The 9th Battalion's troops were also inadequately trained. For example, they did not deploy the smoke canisters mounted on their tanks that would have made the tanks more difficult to target. The tank column called for artillery and air support, but Northern Command denied these requests because of concerns about the possibility of hitting nearby friendly forces.[221] The battalion halted and failed to open the route across the Saluki. Twelve IDF soldiers were killed and 44 were wounded.[222]

---

[219] Katz, "Security and Defense." Here, "Cornet" is the AT-14 Kornet-E ATGM.

[220] Harel and Issacharoff, *34 Days*, p. 225.

[221] David Eshel, "Lebanon 2006," *Tank Magazine*, December 2006; Yaakov Katz, "Wadi Saluki Battle—Microcosm of the War's Mistakes," *The Jerusalem Post*, August 29, 2006.

[222] Katz, "Security and Defense." For overall IDF tank losses, see Alon Ben-David, "Israeli Armour Fails to Protect MBTs from ATGMs," *Jane's Defence Weekly*, October 4, 2006, where Ben-David writes,

> Forty-five per cent of the Israel Defence Force's (IDF's) MBTs [main battle tanks] hit by Hizbullah ATGMs during the fighting were penetrated. Out of 50 IDF Merkava Mk 2, 3 and 4 MBTs hit, 21 were penetrated. Eleven of the incidents resulted in no fatalities while 10 incidents resulted in 23 crew casualties. During the fighting, the IDF encountered a wide variety of Russian- and Iranian-made ATGMs, including the Kornet-E 9P133, claimed to be able to penetrate 1–1.2 m of armour protected by ERA [explosive reactive armor]; the Metis-M 9M131, equipped with a tandem high-explosive anti-tank . . . warhead; the 9K113 Konkurs (AT-5 "Spandrel"); the 9K111 Fagot (AT-4 "Spigot") and the tandem warhead RPG-29 rocket-propelled grenade.

On August 13, the 401st Brigade tried again to force a crossing to the west side of the Saluki. By 0300h it had succeeded, but the offensive was halted by Northern Command and not resumed. The division did not accomplish its missions.[223]

98th Division

IDF paratroopers in this division had begun executing operations on August 8 (prior to implementation of Operation Changing Direction 11) by seizing the village of Debel and staying in houses along the edge of the village. Hezbollah attacked one of these houses with ATGM fire, killing seven and wounding 19.

On August 10, Northern Command ordered the division to take key terrain in and around Rashaf by August 11, a mission accomplished on August 12. However, it did not accomplish the mission of opening a logistical route in the vicinity of the Hubkan Junction.[224]

The final major operation of the division was a helicopter air-landing operation near a village called Yatar. After dropping the paratroopers at the landing site, one of the IAF CH-53 heavy-lift transport helicopters was shot down, and all five crewmembers were killed. The paratroopers were preparing to continue their attack to root out Hezbollah rockets when they were ordered to cease operations.[225] As the Winograd Commission notes, "the progress of the Fire Formation [98th Division] towards its objectives towards Tyre was stopped and not resumed. The formation did not accomplish its missions."[226]

366th Division

The 366th Division's mission was to take the Marj Ayoun area. On August 9, its preliminary attack was stopped by Hezbollah. On August 11–12, the division did not effectively resume its attacks in

---

[223]The Winograd Commission, p. 367.

[224]The Winograd Commission, p. 368.

[225]Harel and Issacharoff, *34 Days*, pp. 232–234; The Winograd Commission, p. 368.

[226]The Winograd Commission, p. 369.

support of Operation Changing Direction 11, and "the result was that the Pillar of Fire Division [366th Division] did not accomplish its missions."[227]

Figure 2.5 presents the final IDF ground positions and provides an assessment of the state of Hezbollah rocket-launching sites at the time when the ceasefire agreement ending the war went in to effect.

Figure 2.5
Final IDF Positions and Hezbollah Rocket-Launching Sites

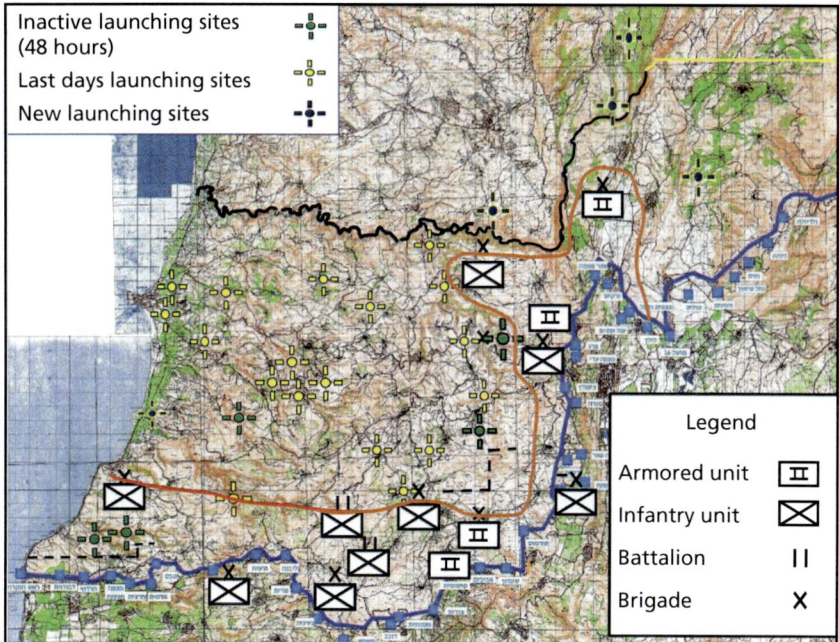

SOURCE: Yechezkel Aguy, "Mechanized Maneuvers in an Anti Tank and Obstacle Saturated Terrain," presentation at the Maneuver in Complex Terrain Conference, Latrun Israel, September 1–3, 2009.
NOTES: The blue line is the border between Israel and Lebanon. The orange line is the depth of the Israeli advance.
RAND MG1085-2.5

---

[227] The Winograd Commission, p. 366.

The End of the War

The ceasefire agreement went into effect at 0800h (Tel Aviv time) on Monday, August 14. Hezbollah moved its forces back from the border, and UN peacekeepers filled in behind them. Overall, approximately 120 IDF soldiers had been killed and more than 1,000 had been wounded in the war; there were also 37 Israeli civilian deaths, mostly as a result of Katyusha fire. Estimates of the number of Lebanese civilians killed range from 900 to 1,100, and the reported number of Hezbollah members killed ranges from 250 to 800.[228] Beyond these figures, however, the war was a disaster for Israel. Just by surviving and showing that it could continuously launch rockets at Israel, Hezbollah was able to claim victory. Indeed, 217 rockets fell on Israel on August 13.[229] The IDF's reputation as a competent military force—a reputation key to Israel's deterrent power—also suffered because of both the indecisive way in which its operations were commanded and its poor performance in the field. These shortcomings were mainly due to Israel's lack of preparation for the war in which it found itself.

Shortfalls Identified in the Second Lebanon War

The abstract for Avi Kober's February 2008 prize-winning essay, "The Israel Defense Forces in the Second Lebanon War: Why the Poor Performance?" succinctly captures the Israeli shortfalls in the Second Lebanon War:

> Whereas in the past, Israeli successes on the battlefield compensated for deterrence and/or early warning failures, in the Second Lebanon War serious problems in Israeli military capabilities and conduct of war were exposed. The article offers explanations for the poor performance of the Israel Defense Forces (IDF): A late perception that it was war; adherence to post-heroic warfare

---

[228]Daniel Byman and Steven Simon, "The No-Win Zone: An After-Action Report from Lebanon," *The National Interest*, No. 86, November/December 2005, p. 58; Cordesman, Sullivan, and Sullivan, *Lessons of the 2006 Israeli-Hezbollah War*, p. 5.

[229]The Winograd Commission, p. 596.

under circumstances that rather required a different approach; the erosion of the IDF's fighting standards due to policing missions which had become its main task since the outbreak of the first Intifada in 1987; false Revolution in Military Affairs inspired concepts; the adoption of the notion of controlling instead of capturing territory; a centralized logistic system; poor generalship; a hesitant and inexperienced political leadership, and IDF dominance in decisions on military matters.[230]

Even more to the point was the testimony of Vice Premier Shimon Peres to the Winograd Commission: "Afterwards I also thought that the IDF was not prepared for this war."[231]

The Winograd Commission was the most authoritative of some 50 investigating teams that examined every aspect of the war.[232] Most of the reports written by these teams are not available to the public, but the interim and final reports of the Winograd Commission were published and translated into English. Additionally, there was a vigorous—and frequently acrimonious—debate (much of it available in English) in the Israeli press among the key senior officers involved in the conduct of the war or in IDF policies before the war who were endeavoring to make sure their side of the story was aired.

Many of the key Israeli shortfalls were discussed earlier in this chapter in the section that discusses the state of the Israeli military in

---

[230]Kober, "The Israel Defense Forces in the Second Lebanon War," p. 3. For information on the breakdown of the centralized logistics system, see Martin van Creveld, "Israel's Lebanese War: A Preliminary Assessment," *The RUSI Journal*, Vol. 151, No. 5, October 2006, pp. 41–42. Van Creveld writes that, in the reserves, "many kinds of equipment such as webbing, bullet-proof vests, and communications gear were in short supply." Furthermore, even food and water were supply issues:

> The part of the logistic system responsible for class 1 supplies did not function properly. It left thousands without either food or water for days on end, forcing them either to rely on handouts from the civilian population—those who had not fled their homes—or else to scavenge for what they could find inside Lebanon itself. In fact, the logistic system was one of the main shortcomings revealed by the war.

[231]Aviram Zino, "I Wouldn't Have Gone to War, Peres Tells Winograd Commission," YNewsNet.com, March 22, 2007.

[232]Harel and Issacharoff, *34 Days*, p. 243.

2006. This section examines several of the more-significant shortfalls that became apparent during the war.

Standoff Fires Are Critical, but Not Sufficient, to Achieving Objectives

The most important observation to be made about the Second Lebanon War is that the IDF's operational concept did not present a pathway for the military to achieve the objectives set by the politicians. The existing concept, which was based largely on standoff fires, did not force Hezbollah to meet Israeli demands. The Winograd Commission was quite explicit in this regard:

> The Air Force filled a vital role in the intense activation of standoff fire. At the same time, it is important to note that this kind of fire, especially from the air—its innovation and scope notwithstanding—has not proved to be a critical weapon under the special circumstances on the Lebanese arena and the fighting against Hizballah, in contrast with the hope and expectation entertained by some members of senior IDF echelons.[233]

Standoff fires were unable to significantly reduce Hezbollah's ability to continue firing short-range rockets against the Israeli homeland.[234] Indeed, although it lauded the IAF's performance in the war, the Winograd Commission raised doubts about the utility of the IDF doctrine of standoff fires, provided principally by the IAF, as a solution to the problem posed by such adversaries as Hezbollah:

> The Israeli Air Force demonstrated exceptional capabilities during this war. Its quality and the fact that it is an important component of Israel's deterrent capability are beyond any doubt. At the same time, it should not be regarded as a "miracle solution" for every wartime need. One should particularly be wary of entertaining excessive expectations with respect to its capabilities in

---

[233] The Winograd Commission, p. 314.

[234] The Winograd Commission, p. 314.

stand-off fire in a confrontation with a well-prepared guerrilla enemy such as Hizballah.[235]

This is a very important statement on two levels. First, the Winograd Commission found that the IDF's perception (articulated in its 2006 operational concept) of the appropriateness of its military capabilities for the full gamut of challenges Israel faced was not adequate.[236] Quite simply, Israel's overall strategic assessment—that the future would be one of deterring high-end threats with the IAF and concentrating the Israeli Army on low-intensity conflicts (such as in Gaza and the West Bank)—was bankrupt. Second, the IDF's operational plans were based on the erroneous assumption that standoff fires, provided principally by air, could either compel a nonstate actor (such as Hezbollah) to stop firing rockets at Israel or force the state (in this case, Lebanon) to make that actor stop the attacks.[237] Again, the Winograd Commission was quite explicit in this regard: "Let us especially point out that the expectation that was entertained by some members of the

---

[235] The Winograd Commission, p. 315.

[236] See The Winograd Commission, p. 262. The IDF's "operational concept" is central to Israeli defense planning and is much broader than the term *operational* implies. The Winograd Commission quotes the following definition of *operational concept* from an IDF anthology on the topic:

> [The operational concept is] the concept regarding the reality in which the IDF operates and regarding the threats with which it is expected to contend. The concept defines the principles for the IDF's operations, including the strategic and other capabilities that are required both in the short-term and in the long-term. It reflects the IDF's view [of] . . . Israel's position in the regional, international, and global environments. It also defines the main threats in the short-term and in the long-term and is affected by the processes that take place in the Israeli society. It also defines the major short-term and long-term threats and is affected by the processes which the Israeli society is undergoing. In essence, it is a framework of ideas, principles, and guideline procedures. The concept outlines and directs the build-up of the force and its activation on the strategic and operative level of the IDF, namely: the General Staff Command, including all its branches, arms, and major operational commands. It lays the foundation for an understanding of the strategic-systemic connections and the linkage between them for the purpose of designing, planning, and managing campaigns, including preparedness for current and future challenges and threats for which it is supposed to prepare.

[237] Catignani, *Israeli Counter-Insurgency and the Intifadas*, p. 188. See also The Winograd Commission, p. 315.

IDF's top echelon that precision stand-off fire that would hit strategic Hizballah targets could decide the battle was wrong."[238]

When standoff fires did not resolve the conflict in Lebanon, the General Staff had no plan beyond an out-of-date Northern Command plan (called Shield of the Land) to fall back on. This plan "was adequate as long as the Syrian Army occupied Lebanon and prior to the great build-up of Hizballah's defense alignment in southern Lebanon."[239] Thus, as the Winograd Commission found,

> on the eve of the Second Lebanon War, neither the General Staff Command nor the Northern Command had a full and validated operative plan; on the other hand, there were several operative plans in different preservation and update levels, where every division individually chose the plan according to which it would prepare. One of the reasons for this was the shunning of a large-scale ground operation.[240]

The Winograd Commission also assessed the consequences of the absence of a coherent plan for Lebanon:

> The absence of plans not only severely encumbered the ability to advance quickly and on the basis of a common language toward operations that were drilled according to the plans while adjusting them to the circumstances, but it also limited the effective readiness on other levels as well.

Thus, when Hezbollah's short-range rocket barrages escalated after the air campaign, there was no accepted Israeli plan to deal with them. The ground war became an act of improvisation by the Israeli Army, which was not trained or prepared for the war it found itself in. This is particularly significant in light of the fact that the IAF had always maintained that it would experience great difficulty—and have a low chance of success—in stopping short-range rockets if its attacks

---

[238] The Winograd Commission, p. 315.

[239] The Winograd Commission, p. 263.

[240] The Winograd Commission, p. 388.

on other, more-identifiable targets in Lebanon did not either change Hezbollah's will to continue firing at Israel or force the Lebanese government to constrain Hezbollah.[241] This finding resulted in one of the Winograd Commission's more damning assessments of the IDF:

> As will be recalled, the Air Force stated from the outset that it would not have great effectiveness in hitting the short-range rocket alignment. This warranted greater General Staff and Air Force focus, both before and during the war, on finding alternative ways to deal with the problem.[242]

The Short-Range Rocket Challenge

The difficulties associated with finding and destroying hidden short-range rockets, particularly ones that do not present a good radar signature that reveals their location, were well understood by the IAF prior to the war. Like most modern air forces, the IAF had concentrated its adaptive efforts on meeting the challenges posed by evolving conventional threats. Figure 2.6 shows which types of targets the IAF has confronted and sought to defeat over time.

During the 1967 War, the principal threat to Israel came from enemy air forces, and the IAF focused on destroying them on the ground. By the 1973 Yom Kippur War, the threat had evolved, and mobile air-defense systems were the primary challenge. Intermediate- and long-range rockets became a palpable threat to Israel in 1991 when Iraqi Scud missiles hit the country during Operation Desert Storm. In 2006's Second Lebanon War, intermediate- and long-range rockets posed a clear threat. As previously discussed, the IDF adapted to these challenges and had, by the time of the war, identified effective solutions for finding and attacking these different sources of threat. What became a problem in 2006—and continues to be a problem to this day—is the issue of countering short-range rockets.

Short-range rockets are hard to find from the air for four principal reasons. First, the rockets are relatively small and, compared with

---

[241] The Winograd Commission, p. 316.

[242] The Winograd Commission, p. 316.

Figure 2.6
The Changing Nature of Israeli Targets

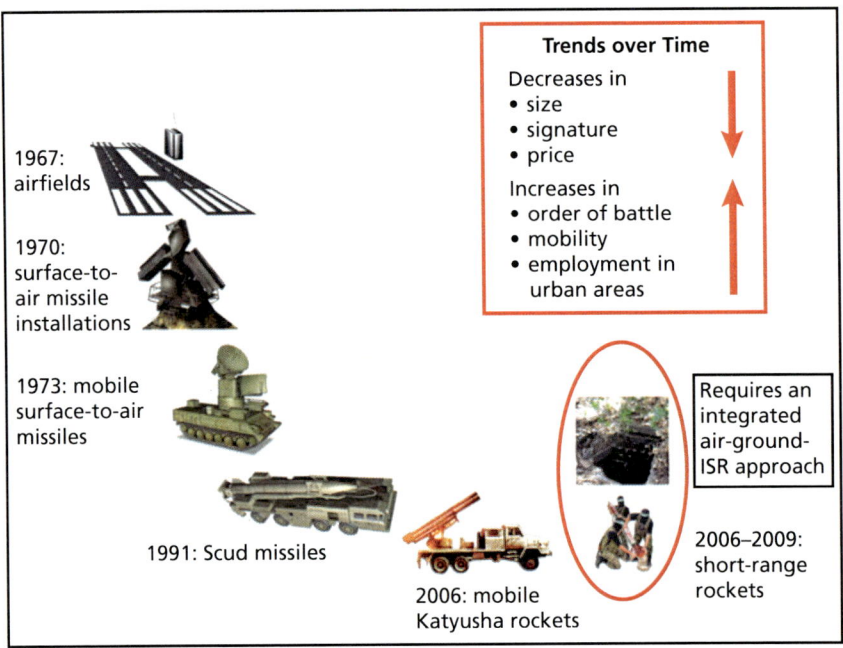

SOURCE: Adapted from material provided to the author by the IAF Doctrine Branch in 2009.

intermediate- and long-range rockets, are easier to hide. Second, these smaller rockets do not need to be attended by crews, and they can be fired remotely or with timers. Third, unlike larger rockets, short-range rockets do not need to be mounted on vehicles, do not create a significant back blast, and do not have to be moved into an open area and prepared for launch by a crew before being fired. (Activities associated with preparing and firing rockets make the longer-range rockets visible to overhead ISR assets.) Fourth, compared with larger rockets, whose physical size when on the launcher and whose launch blast make them easier to find and attack, smaller rockets have a smaller signature. Throughout the Second Lebanon War, despite saturating southern Lebanon with ISR assets and employing multiple attack means, the IDF could not find an air-only solution to the short-range rocket threat. Ground maneuver is necessary to find and neutralize this type of rocket.

There were thousands of short-range rockets in Lebanon and, as discussed in the next chapter, in Gaza. Their small size and low signature, and the fact that they can be transported and smuggled with relative ease, make them both difficult to preempt before they are placed in firing positions and hard to find once they are emplaced. They are an ideal asymmetric weapon. Furthermore, hybrid opponents, such as Hezbollah and Hamas, have a smaller, decentralized order of battle and are able to use complex terrain (especially urban terrain) to good advantage to conceal themselves and to thwart conventional targeting efforts. As the IDF learned in Lebanon, the use of these rockets presents a very different challenge for Israel than that posed by the country's conventional adversaries during past wars or during its recent LIC operations.

When standoff fires did not decide the issue in Lebanon, poorly prepared ground forces were finally brought in. Much of the harshest criticism in the Winograd Commission's final report is aimed specifically at the Israeli Army, but the commission's criticism of the IDF in general is captured in the following quote:

> The impression we gained with respect to the functioning of the Israel Defense Forces, particularly the army, was difficult and bleak. In spite of stubborn, brave fighting by many troops, in regular and reserve service alike, the military as a whole, and the army in particular (as opposed to the Air Force and in partial contrast with the Navy) did not usually succeed in fulfilling most of its missions and challenges. Not only did the Israel Defense Forces not vanquish the Hezbollah, despite its significant quantitative and qualitative edge, but the rocket fire into Israel lasted until the last day of the war.[243]

Criticism of Israeli Army performance also appears in an assessment produced by the Israeli Institute for National Security Studies:

---

[243] The Winograd Commission, pp. 232–233.

During the fighting with Hizbollah, inadequate professionalism of the forces and commanders in some of the combat units was observed. This was the case for regular as well as reserve units. Prior to the war most of the regular forces were engaged in combating Palestinian terror. When they were transferred to Lebanon, they were unfit to conduct combined forces battles integrating infantry, armored, engineering, artillery forces, and other support forces. In some instances, the units lacked both the skills and the necessary organic weapon systems required for this type of fighting. Under these circumstances units found themselves trying to adjust rapidly—often successfully—while engaged in fighting. The professionalism of the reserve troops was not better but for different reasons. It resulted from a years-long process during which the army reserves were neglected. The education and training of the officers were shown to be ineffective. The lack of practical training during reserve duty was evident, as was the lack of cohesion of the units, which had a detrimental effect on their operational capability.[244]

Civil-Military Issues, Unclear Objectives, and a Lack of Jointness

As noted earlier, Israel also experienced problems in the areas of civil-military interaction and decisionmaking. These problems were a serious issue because the discussions between senior Israeli leaders never resulted in the development of a strategy, rooted in political objectives, that could be translated into obtainable—and, in some cases, even understandable—operational and tactical missions. The civilian decisionmakers' lack of military experience may be partly to blame, but the situation is no different in many other countries.

It is also plain that the viability of Israel's air-centric military strategy, which depended on successfully coercing the Lebanese government into acting to control Hezbollah, was highly questionable.

---

[244]Gabriel Siboni, "The Military Campaign in Lebanon," p. 66. See also van Creveld, "Israel's Lebanese War: A Preliminary Assessment," p. 41. The state of the reserves was particularly problematic. Active units experienced difficulty in dealing with the different enemy they found in Hezbollah. In van Creveld's view, "the mobilization process did not come up to expectations. Years had passed since the men had trained together, and a great many of them were out of condition and [had] forgotten how to fight."

For the Israelis in 2006, emulating the air campaign launched against Yugoslavian President Slobodan Milošević during the North Atlantic Treaty Orgnization's 1999 Operation Allied Force, whose purpose was to compel President Milošević to end the abuses Serbs were committing against ethnic Albanians in Kosovo, proved unfeasible.[245] The contextual realities in Kosovo—where President Milošević "had the power to turn on or turn off military action"—were quite different from those in Lebanon,[246] where the prime minister did not have such power. As Neville Parton notes,

> the Lebanese parliament was split almost down the middle, with attitudes towards Syria and Hisbollah marking the dividing line. Of the 128 seats in the parliament, the anti-Syrian camp had a small majority (72 seats)—although this was an alliance grouping, and the Prime Minister's party only had half of these. The rest of the seats belonged to pro-Syrian and pro-Hisbollah factions, who thus held a commanding position within the parliament (and indeed hold two government appointments). . . . [W]hat is clear is that the Prime Minister's authority was quite limited—certainly when it comes to any authority over Hisbollah—which makes the situation very different in terms of the likelihood of a successful coercive approach compared with the situation in Kosovo.[247]

Another factor in Israel's poor performance in Lebanon was the lack of coherence within the IDF itself when it came to issuing clear military orders that framed mission-oriented objectives. Much of the problem within the Israeli military can be traced to the state of flux of the new and evolving operational concept:

---

[245] On Operation Allied Force, see Johnson, *Learning Large Lessons*.

[246] Neville Parton, "Israel's 2006 Campaign in Lebanon: A Failure of Air Power or a Failure of Doctrine?" *Royal Air Force Air Power Review*, Vol. 10, No. 2, Summer 2007, p. 88. For an assessment of why President Milošević responded the way he did during Operation Allied Force, see Stephen T. Hosmer, *The Conflict Over Kosovo: Why Milosevic Decided to Settle When He Did*, Santa Monica, Calif.: RAND Corporation, MR-1351-AF, 2001.

[247] Parton, "Israel's 2006 Campaign in Lebanon," pp. 88–89.

One of the characteristics of the operational concept was the fact that it was formulated on many different levels of discussion, some of them very generalized, and that on key issues, such as the relationship between the use of air fire versus ground moves, it did not explicitly take a position. This presented difficulties in translating the General Staff Command's conception to combat theories and to operational concepts on the fronts and in the corps.[248]

Furthermore, the language used in the concept, derived from SOD, created confusion when orders were being translated into understandable military tasks. The Winograd Commission addressed this issue in its discussion of the 91st Division commander's concept in the July 21 order for the attack on Bint Jbeil (Operation Steel Threads):

> Here we provide an example of two expressions that should have defined the divisional intent: "Systemic *demonstration* on the town of Bint Jbeil, neutralizing, capturing and circumscribing the area, systematically breaking it down by simultaneous, multi-dimensional *swarm melees* . . .," or, in another case: "Widespread low-signature infiltration (wasp cloud), melee—rapid forming up on dominating grounds and forming lethal contact with the built up areas (swarms) while causing shock and awe, freezing the area of operation and transitioning to domination, while systematically breaking down enemy infrastructures in the area."

> There is no dispute that the language of the divisional commander was creative. Some described it as poetic. In part, this was the language that was introduced into use within the new systemic concept and operating concept. Some subordinates were very familiar with it and had no difficulty understanding it. However, not all of the commanders who fought under the command of the divisional commander were his subordinates before the war. It is no surprise that they and their troops had difficulty

---

[248]The Winograd Commission, 263.

in understanding these terms and translating them into operational military language.[249]

Finally, the Winograd Commission pointed out that there were real problems with what the U.S. military would call *jointness*. Some of the issues stemmed from the operating concept, which was not clear about the "relationship between the use of air fire versus ground moves."[250] That said, many of the problems were the result of the IAF and Israeli Army not having trained and planned together. This was particularly evident in IAF close air support for ground operations:

> The Air Force's participation in the ground battle demonstrated many flaws. Basically, they stemmed from the serious flaws in the early planning, preparedness, and training processes involving the cooperation among the IDF branches and from wide gaps in the operational concept. Flaws that surfaced at the beginning of the war were similarly not corrected during the war. These flaws were not local and did not affect merely the Air Force's operations. Their presence is particularly grave in light of the fact that the weak points in the synergy among the forces were known throughout the IDF's wars. In any operation involving the integration of forces, the responsibility for the fact that this integration is not planned, drilled, or assimilated falls on all the parties to that cooperation. The same holds true in our case as well, both in terms of the preparedness and the planning that preceded the war and in terms of the activities performed during the war itself.[251]

---

[249] The Winograd Commission, p. 357 (emphasis in the original). See also Matt M. Matthews, "Interview with BG (Ret.) Shimon Naveh, 1 November 2007," U.S. Army Combined Arms Center Combat Studies Institute, Fort Leavenworth, Kan., undated [2007], p. 2. Brigadier General Gal Hirsch, commander of the 91st Division during the war, was a former OTRI student. During his interview with Matthews, General Naveh identified General Hirsch as "'the most creative thinker, the most subversive thinker and the victim of this entire affair.'"

[250] The Winograd Commission, p. 263.

[251] The Winograd Commission, p. 316.

Hezbollah Presented the IDF with a New Type of Adversary

The Israelis realized in retrospect that Hezbollah constituted a different type of enemy than the IDF had confronted in the West Bank or in Gaza. This realization led to an understanding of the need to prepare in the future for a different kind of war,

> a limited war of a state against a non-state actor operating from the territory of a failed state that does not control its own territory. The non-state player [Hezbollah] fought as a guerilla force, though in some areas it possessed state-like capabilities, acquired from supporting states. For example, Hizbollah had various kinds of guided missiles: anti-tank, anti-aircraft, and land-to-sea missiles as well as assault UAVs, and had the ability to strike deep in Israel's home front.[252]

Hezbollah was an adversary that was neither the low-end nor the high-end threat that Israel had envisioned in its operational concept. Although not ten feet tall, Hezbollah was trained and organized into small units and armed with effective standoff weapons, including ATGMs, RPGs (including highly effective RPG-29s), rockets, mortars, mines, IEDs, UAVs, and MANPADS. Hezbollah also occupied prepared defensive positions in Lebanon's difficult hilly terrain and urban areas. Overall, this kind of adversary presented a qualitative problem that required joint, combined-arms fire and maneuver. In the aftermath of the Second Lebanon War, this is the enemy that the IDF set about preparing to fight.

Hezbollah also required a return to decentralized command and control—aka "mission command"—in tactical operations rather than the reliance on highly centralized control that had characterized Israel's operations to confront the intifada. Indeed, one of the complaints aired to the Winograd Commission by the regional commanders was that the

---

[252] Shlomo Brom, "Political and Military Objectives in a Limited War Against a Guerilla Organization," in Shlomo Brom and Meir Elran, eds., *The Second Lebanon War: Strategic Perspectives*, Tel Aviv: Institute for National Security, 2007, pp. 13–14.

involvement of the General Staff was too deep and even unneces-sary. It inhibited the Regional Command in discharging its duty and led to preoccupation of the General Staff with micromanag-ing the tactical issue at the expense of dealing with the overall strategic view of the campaign.[253]

Indeed, Martin van Creveld notes that "units continued to receive con-tradictory, ever-changing orders" from their higher headquarters.[254]

Finally, the threats posed by rockets created a set of strategic issues within Israel on three levels. First, the fact that the IDF could not stop short-range rockets created a sense of vulnerability in the Israeli popu-lation. After the Second Lebanon War, the Israeli government strength-ened its homeland-defense and civil-defense capabilities to better warn and protect its population from rocket attacks.[255]

Second, the prospect of long-range rockets with larger payloads, particularly if their accuracy is improved, will eventually put all of Israel at risk of attack. Hezbollah's challenge will be finding ways to hide these more-capable systems so that interdicting them, as occurred during the war in 2006, becomes impossible. Hezbollah appears to be working on this project: The IDF believes that Hezbollah is placing rocket-launching sites inside of villages in southern Lebanon. Addi-tionally, the Israelis claim that Hezbollah has acquired "a number of 300-km-range 'Scud' missiles—recently transferred from Syria—and hundreds of Syrian-manufactured, solid-propellant M600 missiles. The M600 is equipped with an inertial navigation system, has a range of 250 km and carries a 500 kg warhead."[256] Consequently, although

---

[253] The Winograd Commission, p. 296. Not surprisingly, the General Staff members, and General Halutz, had a different view: "The General Staff and the IDF Chief of Staff felt that the Regional Command had not met its operational objectives and had not 'come up with the goods.' The IDF Chief of Staff expected the Regional Command to submit proposals and plans for an offensive, and execute them, but this did not happen."

[254] Van Creveld, "Israel's Lebanese War," p. 42.

[255] Author's discussions with IDF officers, Tel Aviv, March 2–5, 2008, and February 9–19, 2009.

[256] Yaakov Katz, "IDF Declassifies Intelligence on Hizbullah's Southern Lebanon Deploy-ment." Katz also notes that

Israel had a number of missile defense efforts under way in 2006 to counter Syrian chemical attacks, the drive to create a comprehensive shield increased in the aftermath of the Second Lebanon War.[257]

Third, Hezbollah's rockets created a military problem for the IDF in that it put IAF air bases at risk. One IDF general officer wrote that,

> during the second Lebanon war, one IAF base and several other important sites (including the IAF central command post in the northern region) came under rocket attack. This necessitates appropriate protection of the air teams and other entities and the development of an operational capability under continuous rocket and SSM [surface-to-surface missile] attacks.[258]

The important issue for the U.S. military in assessing the Second Lebanon War is not determining who won and who lost. Indeed, one can make a very good case that Secretary-General Nasrallah vastly underestimated the Israeli response to the July abductions of the IDF soldiers. As scholars Daniel Byman and Steven Simon point out,

> air strikes and infantry sweeps probably eliminated about half of the longer-range rockets that were not expended, as well as a large number of launchers. Hizballah's elaborate infrastructure

---

> Hizbullah has set up positions, including rocket launchers, arms caches, command-and-control centres and surveillance posts in close to 200 villages in southern Lebanon. . . . [Hizbullah] has also positioned improvised explosive devices (IEDs), some of them weighing up to half a tonne, at the entrance to the villages and along roads it believes the IDF would use in a future ground offensive. . . . The group is believed to have more than 40,000 rockets and missiles compared with 15,000 in 2006, as well as 30,000 guerrilla fighters—20,000 deployed in southern Lebanon—compared with 15,000 in 2006.

[257] Howard Schneider, "Israel's Missile Defense System Is Progressing Steadily," WashingtonPost.com, September 19, 2009. This article also describes the system being put in place in Israel:

> Centered on the Arrow 2 antimissile system, which has been deployed, the project is being extended to include a longer-range Arrow 3, the David's Sling interceptor designed to hit lower- and slower-flying cruise missiles, and the Iron Dome system intended to destroy Grads, Katyushas, Qassams and other shorter-range projectiles fired from the Gaza Strip and southern Lebanon.

[258] Brun, "The Second Lebanon War," p. 39.

in south Lebanon was disrupted, and many of its facilities in the Beirut suburbs were razed. By the time Hizballah was pushing for a cease-fire, which winners do not normally do, its fighters were trapped in a box between the Israeli border, a blockaded coast, blown bridges and roads leading north, and a large IDF force in Marjayoun, poised to march up the Beka to the east.[259]

The problem domestically and internationally was that the results identified by Byman and Simon were obtained in a very ugly fashion. The IDF, for the first time, looked incompetent.

In the aftermath of the Second Lebanon War, the IDF began correcting the deficiencies identified by the Winograd Commission and the other entities that investigated the war. In late 2008, the IDF tested its reforms in the crucible of combat in Gaza, the subject of the next chapter.

---

[259] Byman and Simon, "The No-Win Zone," p. 56.

CHAPTER THREE

Operation Cast Lead

On December 27, 2008, Israel launched military operations (Operation Cast Lead) against Hamas in the Gaza Strip. This confrontation between Israel and Hamas had been brewing since the Israeli withdrawal from Gaza in 2005.

Conflict Backdrop

In August 2005, Israel unilaterally pulled out of Gaza, ending 38 years of occupation. Palestinians in Gaza, the West Bank, and elsewhere believed that the pullout did not actually result in much of a difference in the status of the territory. They argued that Israel still controlled traffic and utilities in and out of Gaza by land and sea and that it refused to allow an airport to be built.[1] The Fatah and Hamas factions jockeyed for position, but Hamas prevailed in elections held in January 2006. In 2007, after a failed coup, Fatah was forced out of Gaza, and Hamas members were similarly ejected from the Fatah-controlled West Bank.[2] Still, Hamas does not have a monopoly over the use of

---

[1] Greg Myre, "As Israelis Pull Out, the Question Lingers: Who'll Control Gaza?" *The New York Times*, September 11, 2005, p. 8.

[2] Council on Foreign Relations, "Hamas," web page, updated August 27, 2009. For an extended discussion of Hamas' takeover of Gaza, see David Rose, "The Gaza Bombshell," *Vanity Fair*, April 2008.

violence in Gaza: The Palestinian Islamic Jihad (PIJ) also carries out "resistance" operations. Hamas and PIJ sometimes even cooperate.[3]

In the period following the Israeli pullout, tensions between Israel and Hamas steadily increased. As Daniel Byman points out, "withdrawal from Gaza did not reduce the power of extremists; instead it led directly to Hamas's takeover of power."[4] Hamas protested Israel's decision to block traffic in and out of Gaza, and Israel complained about rocket and mortar attacks launched from Gaza at Israeli towns in the western Negev. There was something of a lull between June and November 2008, when the two sides limited themselves to relatively low levels of violence under an Egyptian-brokered ceasefire, but tensions persisted.[5]

In 2008, Hamas had stepped up the number of rocket and mortar attacks on Israel. In 2007, Hamas fired 1,115 Qassam rockets and 1,435 mortar rounds into Israel. In 2008, prior to the December initiation of Operation Cast Lead, the number of rockets launched increased to 1,540 and the number of mortar rounds to 1,600. Additionally, 40 of the rockets were the longer-range Grads.[6]

The tenuous ceasefire between Israel and Hamas began unraveling when the Israelis conducted a raid inside the Gaza Strip on November 4. Hamas responded with rockets and mortars, hitting Israel some 200 times between the raid and December 21, despite Israeli warnings that it would launch a significant military operation in response.[7] On December 21, Hamas launched another 70 rockets but, in a statement

---

[3]   Dipak K. Gupta and Kusum Mundra, "Suicide Bombing as a Strategic Weapon," *Terrorism and Political Violence*, Vol. 17, 2005, p. 577.

[4]   Daniel Byman, *A High Price: The Triumphs and Failures of Israeli Counterterrorism*, New York: Oxford University Press, 2011, p. 189.

[5]   Cordesman, "The 'Gaza War': A Strategic Analysis," draft report, February 2, 2009.

[6]   Cordesman, "The 'Gaza War,'" pp. 9, 13. See also United Nations Human Rights Council, *Human Rights in Palestine and Other Occupied Arab Territories: Report of the United Nations Fact Finding Mission on the Gaza Conflict*, A/HRC/12/48, September 15, 2009, pp. 31–32. This report, which is often referred to as the "Goldstone Report" (after the head of the UN Fact Finding Mission, Justice Richard Goldstone), notes that some 8,000 rocket and mortar rounds were fired into southern Israel after 2001.

[7]   Cordesman, "The 'Gaza War,'" p. 9.

that same day, said that the ceasefire (which had expired on December 19) might be renewed "if Israel stopped its aggression . . . and opened up its border crossings."[8] Apparently, as Anthony Cordesman observes, "Hamas did not understand that it was confronting Israel with demands and uses of force where Israel would either have to respond decisively or be seen as having failed to defend itself against the same kind of threat it had faced from the Hezbollah during the fighting in 2006."[9]

On December 27, 2008, Israel executed a massive air attack on Hamas in Gaza. Operation Cast Lead had begun.

The State of the Israeli Military in 2008

The Second Lebanon War was a wake-up call for Israel. The IDF, and particularly the Israeli Army, responded to the many investigations of its performance in Lebanon by going "back to basics." Following the release of the Winograd Commission's final report, Chief of Staff Halutz resigned and was replaced by Lieutenant General Gaby Ashkenazi. In January 2007, the budget cuts from the Kela 2008 defense plan were put on hold, and, in September 2007,

> the Israeli government announced a new defense plan, Teffen 2012. This plan calls for a new emphasis on building up IDF ground forces, including the creation of new infantry brigades. It also foresees adding "hundreds" of Namer heavy armored personnel carriers, several dozen Merkava IV main battle tanks, and a number of tactical UAVs for use at the battalion level.[10]

---

[8] Cordesman, "The 'Gaza War,'" p. 9.

[9] Cordesman, "The 'Gaza War,'" p. 9.

[10] David E. Johnson, Jennifer D. P. Moroney, Roger Cliff, M. Wade Markel, Laurence Smallman, and Michael Spirtas, *Preparing and Training for the Full Spectrum of Military Challenges: Insights from the Experiences of China, France, the United Kingdom, India, and Israel*, Santa Monica, Calif.: RAND Corporation, MG-836-OSD, 2009, pp. 207–208. This information is based on the author's discussions with IDF officers, Latrun, Israel, August 2, 2009. See also Eleanor Keymer, *Jane's World Armies, Issue Twenty-Seven*, Coulsdon, UK: HIS, 2010, pp. 261–262. The Namer is a 60-ton, highly protected armored personnel car-

The renewed procurement of Merkava IV tanks stemmed from Israel's conclusion that the Second Lebanon War suggested that, "properly deployed, the tank can provide its crew with better protection than in the past. The conclusion is that the Israel Defense Forces still requires an annual supply of dozens of advanced tanks in order to replace the older, more vulnerable versions that are still in service."[11] A senior Israeli officer was more blunt: "The FCS [U.S. Army Future Combat Systems] notions concerning intelligence dominance replacing armor are disproved by our lessons. . . . [M]ore-balanced training is not enough. Strykers and MRAPs [mine-resistant ambush-protected vehicles] will not [with]stand a medium-heavy ATGM."[12]

Israel put new emphasis on training in 2007, doubling 2006 training levels.[13] There was broad agreement in Israel that, during the intifadas, the IDF had lost its competence at large-unit maneuver and joint, combined-arms integration. The Winograd Commission was quite specific in this regard:

> A serious shortage in the real drilling of combat capabilities that might be required in a certain arena could become more significant because this is something that cannot so easily be repaired or replenished within a short period of time. Here is it important to point out the fact that since 2000, the IDF as a whole "was sucked into" answering the needs of the military operations on the Palestinian fronts and that since the First Lebanon War in 1982, the IDF had not deployed large ground forces. Operation Defensive Shield was the largest combat operation since 1982, where its scope of operation was smaller than the original Shield of the Land Plan or Mey Merom Plan. This deficiency in holding

---

rier based on the Merkava chassis. It is the IDF solution for transporting soldiers within particularly lethal environments. The IDF is also providing active protection systems for the Merkava and Namer vehicles to defend against ATGMs and RPGs.

[11] Amir Oren, "IDF Girds for Possibility of War with Syria, Hezbollah in 2007," Haaretz. com, last updated June 11, 2006.

[12] Author's email exchange with an IDF officer, December 10, 2010.

[13] Yaakov Katz, "IDF Readying for Gaza Incursion—but Not Yet," *The Jerusalem Post*, September 6, 2007, p. 3.

maneuvers caused the flaws in activating the divisions and the forces and in their maximization, as well as in the effective integration of the various military arms and units.[14]

The comments of Colonel Mickey Edelstein, the up-and-coming commander of the Nahal Brigade in the Second Lebanon War, show the effect that dealing with the intifadas had on preparing for other types of conflicts that required larger-formation fire and maneuver in order to be successful:

> The Nahal Brigade commander Edelstein, who assumed his post a few weeks before the war, had been a brilliant officer in the Ramallah (West Bank) area but had very little experience in leading a large combat formation in more conventional fighting. None of his or 401st commander Kidor's commanders had ever commanded a battalion in full-scale exercises. In April 2007, when the Nahal Brigade conducted its first brigade exercise in seven years (on the Golan Heights), Edelstein admitted to Brigadier General Tzur that this was his first brigade exercise. The division commander brushed it off until Edelstein made himself clearer: This was the first exercise he had *ever* participated in since enlisting 22 years ago.[15]

There were also doctrinal reforms. IDF training, particularly in the Israeli Army, went back to basics, "focusing on bedrock combined-arms fire-and-maneuver tactics and skills, using such terms as *attack* and *defend*" rather than the complex language employed in the 2006 operational concept.[16] There was a clear understanding that consistent

---

[14] The Winograd Commission, p. 387.

[15] Harel and Issacharoff, *34 Days*, p. 226.

[16] Johnson et al., *Preparing and Training for the Full Spectrum of Military Challenges*, p. 208. Indeed, as one Israeli officer told the author, the language may have become too simple in an overreaction to the earlier operational language. He not-too-jokingly said that orders can only have three words in a sentence: e.g., "take the hill" (author's discussion with an IDF officer, Tel Aviv, March 2, 2008). In Avi Gil, "Operation 'Cast Lead'—Gaza: Insight and Lessons Learned from 'Al-Atatra' Battle—the Paratroopers Brigade," briefing, RAND Corporation, Arlington, Va., April 1, 2009, the briefer (then a lieutenant colonel who was with

mission orders were critical and that "tactical forces need: a *defined objective*, clear *missions*, understandable 'in order to' and 'until when.'"[17] There was also a major effort "to introduce a 'new ration between fire and maneuver,' and to develop a new kind of maneuver, different from the classical one[, that] . . . will be adequate to the new challenge—a hybrid threat."[18]

After Lebanon, there was also greater emphasis on professional preparation, with new professional military education courses and training (e.g., brigade and division commander courses) to prepare senior officers for command.[19] Finally, the IDF, and particularly the Israeli Army, returned to the concept of mission command, which is

> a decentralized style of command relying on initiative, the acceptance of responsibility and mutual trust. . . . The underlying requirement [for mission command] is the fundamental responsibility to act, or in certain circumstances to decide not to act, within the framework of the commander's intent.[20]

Although the IDF had long professed a belief in mission command principles, the highly centralized nature of operations during the Second al-Aqsa Intifada—coupled with the issue of the reluctance of senior commanders to trust their subordinates, given the price they believed they would themselves pay for operational errors—had resulted in much-more-centralized control. This was not particularly problematic in the

---

the Paratroopers Brigade during Operation Cast Lead) noted that "attack is attack, defend is defend. Not 'Swarming.'" The Winograd Commission is cautious on this point:

> There is nothing appealing or appropriate in the anti-intellectual tendency among parts of the senior military-command echelon in the IDF. Such a tendency is not conducive to the indispensable deep thinking or strategic conception. It can also lead to a desire to provide superficial correction by pseudo-intellectualism, whose value is doubtful and harmful. (p. 380)

[17] Gil, "Operation 'Cast Lead'—Gaza."

[18] Author's email exchange with an Israeli academic, January 22, 2011.

[19] Author's discussions with IDF officers, Tel Aviv, March 2–5, 2008, and February 8–19, 2009; author's email exchange with an IDF officer, December 10, 2010.

[20] Catignani, *Israeli Counter-Insurgency and the Intifadas*, pp. 57–58.

West Bank and Gaza, but it was a big problem in Lebanon.[21] The Wino-grad Commission, although sympathetic to the fact that "the desired degree of balance between obedience and giving freedom and encouragement for initiative is one of the hardest in the organization of the military," notes the following:

> Generally speaking, in the Lebanon War, we found an excess tendency of commanders and troops to "align" and abstain from action, and we found few cases of persons insisting to act based on their own views, or of offensive initiative and persistence, in the spirit of orders.[22]

Training in the IDF also underwent a radical change. Prior to the Second Lebanon War, roughly 75 percent of training was in LIC, and just 25 percent was in HIC. After Lebanon, the IDF devoted 80 percent of training to high-intensity combined-arms training. In the regular forces, training time was doubled, and combined-arms, live-fire exercises were instituted for brigade combat teams.[23] This is because one of the lessons the IDF learned from the Second Lebanon War is that the "hybrid" threat posed by such adversaries as Hezbollah demands high-end combat skills. Additionally, training and equipping of the reserves has greatly improved.[24]

The Israeli Army and the IAF also markedly increased their cooperation in the realms of ISR, the integration of UAVs, and close air support. The IAF has returned tactical air-control capabilities—which had been removed from the Israeli Army in the years before the Second Lebanon War—to Israeli Army brigades.[25] However, there are staff-

---

[21] Author's discussions with IDF officers, Tel Aviv, March 2–5, 2008, and February 8–19, 2009; Eitan Shamir and Uzi Ben-Shalom, "Mission Command Philosophy from Theory to Practice: The IDF Case," in Military Psychology Center, Ground Forces (IDF), "Abstracts: The 3rd International Military Psychology Conference in Israel," undated [February 2008], p. 61.

[22] The Winograd Commission, p. 278.

[23] Author's email exchange with an IDF officer, December 10, 2010.

[24] Author's discussions with IDF officers, Tel Aviv, March 2–5, 2008, and February 8–19, 2009.

[25] See Johnson et al., *Preparing and Training for the Full Spectrum of Military Challenges*, p. 208.

ing issues associated with providing this capability. Generally, the IAF uses fighter pilots as tactical air controllers, but, given the demand, the IAF is now using reserve fighter pilots and considering using UAV and attack helicopter pilots in these positions. Additionally, there is a center that prepares IAF officers to serve in Israeli Army units.[26]

By the time of Operation Cast Lead in December 2008, many of the reforms just described were in various stages of progress, and many are still ongoing as of this writing. The units allocated to Operation Cast Lead (e.g., Paratroopers [35th Brigade], Golani [1st Brigade], Givati [84th Brigade], and Tracks of Iron [401st Armor]) were among the best in the IDF. These units had trained for and had rehearsed their missions in Operation Cast Lead for some time before the operation.[27]

The Challenges Posed by the Terrain and by Hamas

In Operation Cast Lead, the IDF faced different challenges than it had confronted in the Second Lebanon War.

Terrain

The Gaza Strip is about 41 kilometers (25 miles) long and between 6 and 12 km (4 and 7.5 miles) wide, with a total area of 360 km^2 (139 square miles; about twice the size of Washington, D.C.). It is much smaller than Lebanon. Gaza's population is estimated at 1.5 million people.[28] Most rockets and mortars are fired from the north, which is closer to Israeli population centers in the western Negev. The southern area, which shares a border with Egypt, is well known for its underground activities and for being the site of weapon and money smuggling from Egypt. Central Gaza is mainly agricultural, but it also hosts some refugee camps. The entire strip is easily within reach of IDF fires. Compared with the complex terrain of southern Lebanon, the Gaza Strip is flat, sparsely vegetated, and exposed (see the map in Figure 3.1);

---

[26] Author's discussions with IDF officers, Tel Aviv, March 2–5, 2008, and February 8–19, 2009.

[27] Author's discussions with IDF officers, Tel Aviv, February 8–19 2009.

[28] Central Intelligence Agency, "Gaza Strip," *The World Factbook*, no date available.

it is almost completely surrounded by Israel and by the Mediterranean Sea, which Israel actively and closely patrols.[29]

Hamas' Preparation of the Battlefield

The Gaza Strip's dense population and urban areas, particularly in Gaza City, make it difficult to attack without inflicting harm on civilians; it truly creates a situation of "war amongst the people."[30] Hamas took advantage of the urban areas, using buildings to conceal military materiel and shooting rockets into Israel from areas populated with civilians.[31]

Hamas also developed an extensive network of tunnels along the border with Egypt to smuggle weapons, money, and other goods into and out of Gaza. It had dug other tunnels to use as bunkers and traps to kidnap IDF personnel (e.g., by placing dummies wearing Hamas military uniforms in houses that were rigged with explosives) and to conceal rockets and their launchers.[32] Finally, it had prepared fighting positions on the outskirts of Gaza City where it wanted to fight any Israeli ground intrusion.[33]

---

[29] Author's discussions with IDF officers, Tel Aviv, February 10–12, 2009, and September 2–10, 2009; author's discussions with IDF officers, Washington, D.C., February 26, 2009, and April 1, 2009.

[30] The phrase *amongst the people* is from a chapter titled "War Amongst the People" in Rupert Smith, *The Utility of Force: The Art of War in the Modern World*, New York: Vintage Books, 2005.

[31] Steven Erlanger, "A Gaza War Full of Traps and Trickery," NYTimes.com, January 11, 2009. See also Israel Ministry of Foreign Affairs, "The Operation in Gaza—Factual and Legal Aspects," web page, July 29, 2009; Israel Ministry of Foreign Affairs, *Initial Response to Report of the Fact Finding Mission on Gaza Established Pursuant to Resolution S-9/1 of the Human Rights Council*, September 24, 2009; and United Nations Human Rights Council, *Human Rights in Palestine and Other Occupied Arab Territories.* The UN report is highly critical of Israel's operations in Gaza, contending that Israel used disproportionate force and caused great damage to and destruction of civilian property and infrastructure and great suffering to civilian populations (p. 21). The two Israel Ministry of Foreign Affairs sources rebut the UN report.

[32] IslamicNews.net, "Khottat al-Moqawama fi harb Gaza . . . Istrategiya Mahaliya [The Resistance Plan During the Gaza War . . . a Local (i.e., National) Strategy]," undated.

[33] Author's discussions with IDF officers, Tel Aviv, February 10–12, 2009, and September 2–10, 2009; author's discussions with IDF officers, Washington, D.C., February 26,

Figure 3.1
The Gaza Strip

SOURCE: Adapted from United Nations Office for the Coordination of Humanitarian Affairs, "Gaza Situation Map," May 2006.
NOTE: The Philadelphi Corridor is roughly delineated by the gold line along Israel's border with Egypt.
RAND MG1085-3.1

2009, and April 1, 2009.

Hamas: A Hybrid Opponent, but Not Hezbollah

Hamas is Sunni and the largest of the Palestinian, Islamic fundamentalist militant organizations.[34] Formed in 1987 at the beginning of the first intifada, Hamas has become a key player in Middle Eastern politics. Hamas, an offshoot of the Egyptian Muslim Brotherhood, "combines Palestinian nationalism with Islamic fundamentalism. Its founding charter commits the group to the destruction of Israel, the replacement of the PA [Palestinian Authority] with an Islamist state on the West Bank and Gaza, and to raising 'the banner of Allah over every inch of Palestine.'"[35] Article 8 of Hamas' charter is "The Motto of the Islamic Resistance Movement":

> God is its goal;
> The messenger is its Leader.
> The Quran is its Constitution.
> Jihad is its methodology, and
> Death for the sake of God is its most coveted desire.[36]

An Israeli missile killed Hamas' founder and leader, Sheikh Ahmad Yassin, in March 2004. Ismail Haniyeh, the Gaza government's prime minister, is the organization's senior figure in Gaza. Chairman Khaled Meshaal, Hamas' leader, lives in Damascus, Syria.[37]

Hamas broke from the nonviolent activism of the Muslim Brotherhood in 1988 when it committed itself to the objective of driving Israeli forces out of the occupied territories and placing itself at the forefront of Palestinian resistance to the Israeli occupation. Hamas tactics include suicide bombings and rocket attacks against civilians. Its founding charter calls for the destruction of Israel and for the establish-

---

[34] Jonathan Schanzer, *HAMAS vs. FATAH: The Struggle for Palestine*, New York: Palgrave Macmillan, 2008. *Hamas* is an acronym for the Arabic phrase "Haraka al-Muqawama al-Islamiyya [The Islamic Resistance Movement]"; the word *hamas* also means "zeal" in Arabic.

[35] Council on Foreign Relations, "Hamas."

[36] Khaled Hroub, *HAMAS: Political Thought and Practice*, Washington, D.C.: Institute for Palestine Studies, p. 272.

[37] Council on Foreign Relations, "Hamas."

ment of an Islamic state in all of historic Palestine, including Israel. The United States, the European Union, and Israel have designated Hamas as a terrorist organization.[38]

In January 2006, Hamas defeated Fatah, the party of Palestinian Liberation Organization (PLO) President Mahmoud Abbas, in elections for the Palestinian Legislative Council, which is the legislature of the Palestinian National Authority. In summer 2007, tensions between Hamas and Fatah erupted, and

> Hamas routed Fatah supporters, killing many and sending others fleeing to the West Bank. The result was a de facto geographic division of Palestinian-held territory, with Hamas holding sway in Gaza and Fatah maintaining the internationally recognized Palestinian Authority government in the West Bank town of Ramallah.[39]

Since its inception, Hamas has divided its activities into three main spheres of operation: (1) a political section involved in Palestinian politics; (2) a social section (modeled on those of the Muslim Brotherhood and Hezbollah) that provides basic social services, such as hospitals, schools, and religious institutions, to its constituencies; and (3) a militant section, represented by its paramilitary wing, the Izzedine al-Qassam Brigades, which engage in acts of terror against Israelis and also participate in conflict against other Palestinian factions.[40]

The Izzedine al-Qassam Brigades

The Izzedine al-Qassam Brigades are the main militant-terrorist organization in the Gaza Strip. They are organized into several semimilitary echelons, including brigades, battalions, companies, platoons, and teams, and they have more than 10,000 operatives.[41] These operatives

---

[38] Council on Foreign Relations, "Hamas."

[39] Council on Foreign Relations, "Hamas." For a detailed discussion of the relationship between Hamas and Fatah, see Schanzer, *HAMAS vs. FATAH.*

[40] Council on Foreign Relations, "Hamas."

[41] Intelligence and Terrorism Information Center at the Israel Intelligence Heritage & Commemoration Center, *Hamas's Military Buildup in the Gaza Strip*, April 8, 2008.

reinforce the "regular" hard core of several hundred skilled fighters, who are also supplemented by other fighters.[42] The force is organized into four regional sectors: Northern sector (one brigade), Gaza City sector (two brigades), Central sector (one brigade), and Southern sector (two brigades). The organization is currently commanded by Ahmed Ja'abari and Muhammad Deif; the latter is rumored to have been wounded in a 2006 Israeli attack.[43]

Hamas' skilled fighters are mainly trained in Lebanon by Hezbollah, Iran, and Syria. Hezbollah provides specialized training in the use of standoff weapons, such as ATGMs, MANPADS, and rockets. Most fighters receive at least six months of basic military training involving rocket and mortar launchers. Al-Qassam fighters also participate "in ideological classes held in mosques."[44] The fact that Shia Iran supports Sunni Hamas is important and reminds one of the continued relevance of the age-old adage, "The enemy of my enemy is my friend."

Internal Security Forces

Hamas has also developed its internal security forces, which are mainly in charge of controlling the Gaza Strip and protecting Hamas leaders. In addition to a police force, there is a rapid-response unit and a security and protection unit; there are also units for national security, internal security, coastal security, and civil defense.[45] These units and their members can be called to join the al-Qassam Brigades in case of a war or other emergency.[46]

---

[42] Intelligence and Terrorism Information Center at the Israel Intelligence Heritage & Commemoration Center, *Hamas's Military Buildup in the Gaza Strip*, p. 10.

[43] Intelligence and Terrorism Information Center at the Israel Intelligence Heritage & Commemoration Center, *Hamas's Military Buildup in the Gaza Strip*, pp. 10–12.

[44] Amos Harel, Avi Issacharoff, and Haaretz Correspondent, "Analysis: A Hard Look at Hamas' Capabilities," Haaretz.com, December 26, 2008; author's discussions with IDF officers, Tel Aviv, February 10–12, 2009, and September 2–10, 2009; author's discussions with IDF officers, Washington, D.C., February 26, 2009, and April 1, 2009.

[45] For additional details and description, see Intelligence and Terrorism Information Center at the Israel Intelligence Heritage & Commemoration Center, *Hamas's Military Buildup in the Gaza Strip*.

[46] Intelligence and Terrorism Information Center at the Israel Intelligence Heritage & Commemoration Center, *Hamas's Military Buildup in the Gaza Strip*, p. 46.

After the IDF's unilateral withdrawal from the Gaza strip in 2005, and especially after the Second Lebanon War between Hezbollah and Israel, Hamas engaged in an aggressive military buildup.[47] Prior to the Israeli pullout, Hamas and other militant organizations in the Gaza Strip and the West Bank were under intense Israeli surveillance, and their actions were mainly limited to random mortar attacks and suicide operations against Israel. After the IDF's withdrawal, Hamas reorganized its al-Qassam Brigades into semimilitary formations, improved its command and control capabilities, and increased the strength of its force to around 15,000. It trained some of its forces in Gaza but, as noted earlier, also sent some to Syria, Iran, and Lebanon. Hamas' fighters are not believed to be as well trained as Hezbollah's.[48]

Hamas' Weapons

Prior to Operation Cast Lead, Hamas had procured weapons and ammunition with the help of Hezbollah, Iran, and Syria. It had also manufactured Qassam rockets and a variety of IEDs. Hamas focused on fielding weapons that were suitable for guerilla warfare. Getting weapons into Gaza is difficult because they must enter either through tunnels from Egypt or over the Mediterranean shore, which is closely monitored by the IDF. Moving large rockets via the tunnels presents a real challenge, and, according to reports, the larger, 122-mm Grad rockets had to be disassembled into four pieces to enable their transit. The indigenously produced Qassam rockets have limited accuracy, poor reliability, and short shelf lives. Hamas uses no guided rockets; like Hezbollah, its goal is simply to hit Israel.[49]

---

[47] For details on Hamas' military buildup, see Intelligence and Terrorism Information Center at the Israel Intelligence Heritage & Commemoration Center, *Hamas's Military Buildup in the Gaza Strip.*

[48] Matthews, "Hard Lessons Learned," p. 25.

[49] Alon Ben-David, "Iranian Influence Looms as Fragile Gaza Ceasefire Holds," *Jane's Defence Weekly*, January 22, 2009; Yiftah Shapir, "Hamas' Weapons," *Strategic Assessment*, Vol. 11, No. 4, February 2009. Shapir provides a very detailed discussion of the rockets available to Hamas. See also Harel, Issacharoff, and Haaretz Correspondent, "Analysis: A Hard Look at Hamas' Capabilities," which notes that "Hamas has also made significant gains in manufacturing its own rockets. It has learned to create Ammonium Perchlorate Compound,

On the eve of Operation Cast Lead, Hamas possessed some 3,000 rockets. Most were the indigenously produced variants of Qassams, with ranges of between 3 km and 17 km, but it also had several dozen intermediate-range Grad and WS-1E 122-mm rockets, whose maximum ranges are 20 km and 45 km, respectively.[50] The characteristics of these rockets are listed in Table 3.1.

Hamas also has mortars (81 mm, 82 mm, and 120 mm) whose ranges are between 9 km and 13 km. It was, however, the introduction of the Grads and especially the WS-1E rockets that was of particular concern to the Israelis because these rockets opened much of southern Israel to attack, including the towns of Ashquelon, Ashdod, and Rehovot, and put more than a million Israeli citizens within range of Hamas rocket fire.[51] The longer-range rockets also brought key infrastructure (e.g., ports, a desalination plant, a major electric power plant) within range.[52] Indeed, in the view of many IDF officers, the short-range Qassams cannot be stopped through military action but, because of their small payloads and limited range, are more of a nuisance than a threat. The Grads and WS-1E's are a different matter entirely, given their ranges, reliability, and payloads.[53]

Finally, Hamas also possessed ATGMs (including Sagger missiles), RPGs (including RPG-29s), a small number of SA-7 MANPADS, and an arsenal of small arms, machine guns, sniper rifles, mines and IEDs (including IEDs with explosively formed projectiles), munitions for suicide bombers, and some small boats.[54]

---

an advanced rocket propellant that in addition to extending the Qassam's range beyond 20 kilometers also—and more importantly—increases the rocket's shelf life."

[50] Ben-David, "Iranian Influence Looms as Fragile Gaza Ceasefire Holds"; Shapir, "Hamas' Weapons."

[51] Gil, "Operation 'Cast Lead'—Gaza."

[52] Cordesman, "The 'Gaza War,'" p. 8.

[53] Author's discussions with IDF officers, Tel Aviv, February 10–12, 2009, and September 2–10, 2009; author's discussions with IDF officers, Washington, D.C., February 26, 2009, and April 1, 2009.

[54] For additional details and description, see Intelligence and Terrorism Information Center at the Israel Intelligence Heritage & Commemoration Center, *Hamas's Military Buildup in the Gaza Strip.*

Table 3.1
Hamas Rockets

	Qassam-1	Qassam-2	Qassam-3	Qassam-4	Grad	WS-1E
Length	~80 cm	~180 cm	~200 cm	Unknown	283 cm	294 cm
Diameter	~60 mm	~150 mm	~170 mm	Unknown	122 mm	122 mm
Weight	~5.5 kg	~32 kg	~90 kg	Unknown	72 kg	74 kg
Payload	0.5 kg	5–9 kg	10–20 kg	Unknown	18 kg	18–22 kg
Maximum range	3–4.5 km	8–9.5 km	10–12 km	15–17 km	18–20 km	34–45 km

SOURCE: GlobalSecurity.org, "HAMAS Rockets," web page, date not available. See also "Palestinian Weapons Deployed Against Israel During Operation Cast Lead," *Journal of Palestine Studies*, Vol. 38, No. 3, Spring 2009, pp. 192–200, which notes that, "in February 2007, Hamas confirmed that a Qassam 4 with a range of 15–17 km was in the development stage, but there is no evidence that one has ever been fired, including during . . . [Operation Cast Lead]."

How Operation Cast Lead Unfolded

According to an official IDF spokesman, Operation Cast Lead was launched in response to "the continuation of terror activity by Hamas terror organization from the Gaza Strip, and the duration of rocket launching and targeting Israeli civilians."[55] What soon became apparent was that Israel had entered into this operation with much more preparation than it had in Lebanon in 2006. Operation Cast Lead's objectives were clear from the outset and could be translated into military action. The overall objective was to "create conditions for a better security situation in southern Israel." To accomplish this, Israel planned to inflict "severe damage" on Hamas, decrease the number of terror and rocket attacks originating in Gaza, and increase Israel's military deterrent, all while "minimizing collateral damage and avoiding escalation in other fronts." The desired end state was the long-term "cessation of terror attacks" (including rocket attacks) originating in the Gaza strip.[56]

The military aspects of the campaign were directed by the IDF's Southern Command in coordination with the General Staff in Tel Aviv. This was a distinct change from the Second Lebanon War, when the General Staff retained considerable control, which caused serious issues between it and Northern Command, the headquarters with nominal control over the war.[57]

The Israelis had planned and practiced Operation Cast Lead for months before the first air attack. They had also spent the intervening years since the Second Lebanon War correcting many of the deficiencies in planning, joint operations, training, and operational security

---

[55] Israel Defense Forces, "Operation Cast Lead Against Terror Infrastructure in Gaza Continues," press release, December 27, 2008.

[56] Author's discussions with IDF officers, Tel Aviv, February 10–12, 2009, and September 2–10, 2009; author's discussions with IDF officers, Washington, D.C., February 26, 2009, and April 1, 2009.

[57] Author's discussions with IDF officers, Tel Aviv, February 10–12, 2009, and September 2–10, 2009; author's discussions with IDF officers, Washington, D.C., February 26, 2009, and April 1, 2009; Cordesman, "The 'Gaza War,'" p. 39.

identified in Lebanon.[58] Finally, the IDF, and particularly the Israeli Army, approached Operation Cast Lead with a different mindset than that which prevailed during the Second Lebanon War. In the earlier conflict, the IDF initially used tactics that had been conditioned by years of LIC. In Operation Cast Lead, however, the mentality was different: "It's better to conduct a Special Operation like a full scale war, rather than to conduct a war like a Special Operation."[59]

The Air Campaign

The years Israel had spent in developing intelligence in Gaza resulted in a target set, managed by Southern Command, that included some "603 major targets." This set "treated virtually every known Hamas location or residence as a potential area of operations and part of the Hamas leadership and military infrastructure."[60] This effort had created a

> "mosaic" of targets over a period of several years where highly detailed imagery and COMINT [communications intelligence] were supplemented by effective HUMINT to create a remarkably accurate picture of Hamas targets in Gaza that . . . [the IDF] constantly updated on a near real time basis. The IDF also cooperated directly with Israel's civil intelligence branch—the Shin Bet—in developing its targets for the first time, which gave the IDF improved access to Palestinian HUMINT as well as technical intelligence.[61]

By all accounts, Israel caught Hamas by surprise. This was enabled by

---

[58] Author's discussions with IDF officers, Tel Aviv, February 10–12, 2009, and September 2–10, 2009; author's discussions with IDF officers, Washington, D.C., February 26, 2009, and April 1, 2009.

[59] Gil, "Operation 'Cast Lead'—Gaza."

[60] Cordesman, "The 'Gaza War,'" p. 16.

[61] Cordesman, "The 'Gaza War,'" p. 15.

high levels of secrecy compartmentation to ensure that its [Israel's] war plans did not leak. It prepared a campaign [that] . . . ensured that there would be minimal media coverage in an area where virtually any image or report could aid Hamas. It made sure that its forces did not bring cell phones into the area. Hezbollah's ability to listen to, and locate, cell phone traffic had been a major problem in the fighting with Hezbollah.[62]

Furthermore, a carefully developed deception plan helped ensure that Hamas had no warning of the impending attacks. Consequently, Hamas "did not disperse its leaders or key assets."[63]

In the first strike, launched on December 27, 88 IAF strike aircraft hit 100 preplanned targets in 220 seconds.[64] These aircraft approached Gaza from the Mediterranean along commercial flight paths to further deceive Hamas.[65] This was the largest IAF strike since the 1967 Six Day War. These first-day targets "included command centers, training camps, various Hamas installations, rocket manufacturing facilities and storage warehouses."[66] In this initial phase of the operation, the IDF sought to surprise Hamas with a large standoff attack, destroy hundreds of targets, and prepare the battle space for IDF ground forces.[67] Over the next several days, the IAF continued to bomb targets in Gaza, including hundreds of tunnels, many of which were in the southern part of the territory, near Egypt.[68]

---

[62] Cordesman, "The 'Gaza War,'" p. 15.

[63] Cordesman, "The 'Gaza War,'" p. 15.

[64] Barbara Opall-Rome, "In Gaza, Both Sides Reveal New Gear," DefenseNews.com, January 5, 2009.

[65] Cordesman, "The 'Gaza War,'" p. 15.

[66] Israel Defense Forces, "Operation Cast Lead Against Terror Infrastructure in Gaza Continues."

[67] Gil, "Operation 'Cast Lead'—Gaza."

[68] Cordesman, "The 'Gaza War,'" p. 20. Cordesman includes in his report the daily press statements issued by the IDF on the progress of Operation Cast Lead (pp. 20–27 and pp. 42–57). These statements are also available at Israel Defense Forces, "IDF Press Releases," web page, undated.

The air effort was significant. During the course of the operation, the IAF flew 5,650 sorties for a total of 20,650 flight hours. There were 1,700 fixed-wing attacks and 1,150 attack-helicopter attacks against a total of 3,430 targets. Of the 5,500 munitions employed, 81 percent were precision guided, compared with just 36 percent during the Second Lebanon War.[69]

The Ground Campaign

Unlike the Second Lebanon War, Operation Cast Lead included, from the very beginning, plans for a ground campaign and reserve mobilization. These activities were deliberately incorporated into the operation a week after the air campaign began rather than undertaken as improvisational escalations after the war was under way.[70] It was important to Israel to demonstrate—both to its adversaries and its own population—the competence of its ground forces and its willingness to use them. This was particularly true in the case of calling up reserves, an act that is very expensive and inherently disruptive to day-to-day life in Israel.[71]

As previously noted, Southern Command, commanded by Major General Yoav Galant, was in charge of the overall operation. Its subordinate command, the Gaza Territorial Division (with two regional brigades), commanded by Brigadier General Eyal Eisenberg, was the tactical headquarters that directed the Paratroopers Brigade, the Givati Brigade, the Golani Brigade, the 401st "Tracks of Iron" Armored Brigade, and several IDF reserve brigades. Figure 3.2 shows the broad features of the ground campaign.

---

[69] Dan Harel, "81% of the Fire—Precision Guided Munitions," *Israel Defense*, Vol. 1, February–March 2011, p. 20. Major General Dan Harel was the IDF deputy chief of staff during Operation Cast Lead.

[70] Author's discussions with IDF officers, Tel Aviv, February 10–12, 2009, and September 2–10, 2009; author's discussions with IDF officers, Washington, D.C., February 26, 2009, and April 1, 2009.

[71] Author's discussions with IDF officers, Tel Aviv, February 10–12, 2009, and September 2–10, 2009; author's discussions with IDF officers, Washington, D.C., February 26, 2009, and April 1, 2009.

Figure 3.2
The IDF Ground Campaign in Operation Cast Lead

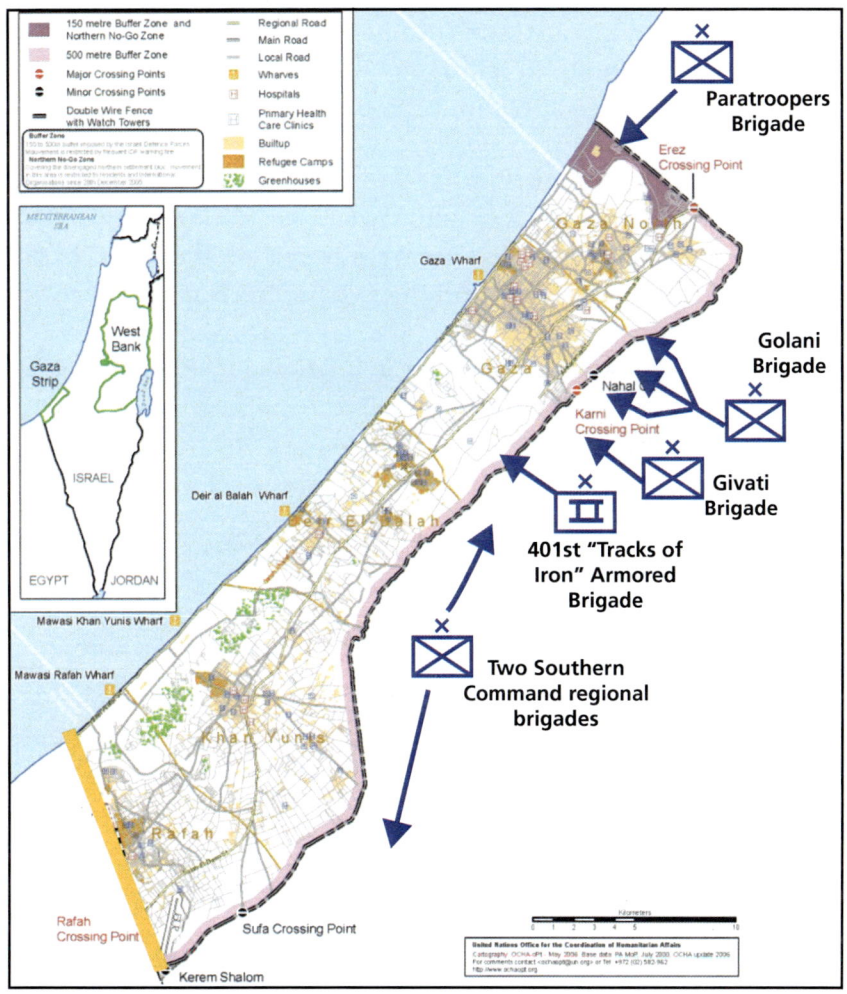

SOURCE: United Nations Office for the Coordination of Humanitarian Affairs, "Gaza Situation Map."
NOTE: The disposition of forces is based on information from Gil, "Operation 'Cast Lead'—Gaza."

The ground phase of Operation Cast Lead sought to encircle Gaza City and push rocket launchers and Hamas fighters out of their prepared positions and into Gaza City, thus reducing Hamas' capabilities and separating Gaza City from the rest of the Gaza Strip.[72] The Paratroopers Brigade advanced south into northern Gaza. Golani Brigade forces entered Gaza on a southwesterly track in a three-pronged attack, and Givati troops engaged northern Gaza while moving northwest. The Tracks of Iron Tank Brigade pushed to the Netzarim junction to block the roads from Rafah and Khan Yunis to Gaza City, thereby cutting off Hamas' supply lines from the tunnels in the south. Northern Command's two regional brigades controlled Israel's border with Gaza.[73]

In Operation Cast Lead, unlike in Lebanon in 2006, the IDF had a better understanding (enabled by better intelligence) of the enemy's "scheme of maneuver." The Hamas scheme of maneuver was based on the following activities:

- using standoff tactics "with snipers, indirect fire[,] and ATGMs to kill Israeli soldiers and lower morale"
- using suicide bombers "to breach IDF forward operating bases and defensive perimeters"
- emplacing IEDs in "roads, houses, schools, mosques, etc."
- using tunnels to create "maneuver space" and for kidnapping Israeli soldiers, activating IEDs, and storing weapons
- executing "'bait and ambush attacks' to kill Israeli soldiers"
- continuing to "launch rockets into southern Israel cities to delegitimize the IDF offensive and boost Palestinian morale."[74]

IDF artillery and air strikes "paved the way by fire" for ground maneuver, hitting Hamas positions and detonating mines and IEDs. Assuming that Hamas had mined and prepared roads as ambush sites,

---

[72] Gil, "Operation 'Cast Lead'—Gaza."

[73] Hanan Greenberg, "Brigade Commander: Hamas Will Draw Lessons, Grow Stronger," YNetNews.com, last updated January 23, 2009.

[74] Gil, "Operation 'Cast Lead'—Gaza."

IDF engineers used armored D-9 bulldozers and explosive line charges to clear remaining IEDs and mines and to create routes for IDF ground forces that avoided these roads.[75] Each brigade had dedicated air assets and controllers, and IAF fixed-wing aircraft continued to bomb the Philadelphi Corridor and the tunnel complexes in southern Gaza.[76]

The firepower that preceded the ground attack, and the rapidity of the maneuver, surprised Hamas. It was driven from its generally well-organized and well-prepared positions back to improvised positions. Thus, what the IDF faced was a largely ruptured defense with an opponent knocked off balance.[77] Furthermore, the al-Qassam Brigades "displayed a low level of combat proficiency."[78]

In the north, Israeli ground forces quickly encircled Gaza City. The IDF attacked mostly at night in order to take advantage of the lack of night-vision capability among Hamas forces.[79] Although the IDF did not put a complete halt to rocket launches from Gaza into Israel, it was able to decrease the number of launches. There was an average

---

[75] Author's discussions with IDF officers, Tel Aviv, February 10–12, 2009, and September 2–10, 2009; author's discussions with IDF officers, Washington, D.C., February 26, 2009, and April 1, 2009; Matthews, "Hard Lessons Learned," p. 3.

[76] See Moshe Hirsch, "Legislation Note: Treaty-Making Power: Approval of the Israel-Egypt 'Philadelphi Accord' by the Knesset," *Israel Law Review*, Winter 2006. Hirsch explains that

> the Philadelphi route [corridor] is a fourteen kilometer strip running along the border between Egypt and the Gaza Strip, from the Mediterranean Sea coastline to the point opposite Kerem Shalom (the Northern border crossing between Israel and Egypt). Following Israel's decision to withdraw its forces from Gaza Strip ("the Disengagement Plan"), Israeli officials were concerned that Palestinians would smuggle in weapons from neighboring Egypt to Palestinian terrorists. While under the original Disengagement Plan Israeli forces were to be deployed along the Philadelphi route, [Israel's] . . . interest to free itself of the responsibility for the Gaza Strip led it to conclude the Philadelphi Accord with Egypt. Under the Philadelphi Accord, Egypt was authorized to deploy 750 border guards opposite the Philadelphi route to patrol the border on its side, and prevent arms smuggling from Sinai (Egypt) to the Gaza Strip, infiltration and other criminal activity.

[77] Gil, "Operation 'Cast Lead'—Gaza."

[78] Yoram Cohen and Jeffrey White, *Hamas in Combat: The Military Performance of the Palestinian Islamic Resistance Movement*, Policy Focus No. 97, Washington, D.C.: The Washington Institute for Near East Policy, 2009, p. 15.

[79] Cordesman, "The 'Gaza War,'" p. 41; Matthews, "Hard Lessons Learned," p. 31.

of 50 launches a day during the first week of Operation Cast Lead but only an average of four per day during the last week of the operation.[80]

Hamas' Military Strategy During Operation Cast Lead

During Operation Cast Lead, early Israeli ground maneuvers pushed Hamas out of many of its prepared fighting positions and thereby negated much of its military potential. Hamas retreated to urban areas and generally did not engage the IDF in open areas. Instead, it sought to channel the fighting into densely populated and built-up areas. Like Hezbollah, its objective was to exhaust and attrite IDF forces and to maintain its attacks on Israeli towns with rockets and mortars.[81] At the same time, it tried to preserve the survival of its operatives, political leaders, commanders, and military infrastructure. Hamas viewed its leaders' survival as both an important mission for its security apparatus and a way to thwart Israel.[82] In addition to preserving and protecting the organization's leadership, Hamas forces sought to protect and retain their principal bargaining chip—captive Israeli soldier Gilad Shalit. They also sought, unsuccessfully, to kidnap other Israeli soldiers during the operation.[83]

Hamas anticipated that the IDF would isolate Hamas units and disrupt the organization's communications. It operated in a decentralized manner, delegating as much authority as possible to personnel in the field and directing local commanders to conduct their own operations without referring to the central command. As in Hezbollah, every unit in Hamas had sufficient provisions, ammunition, and weapons to allow it to survive and fight unsupported for extended peri-

---

[80] Author's discussion with an IDF officer, Washington, D.C., February 26, 2009; Dan Harel, "The Fire Delivery Concept in Operation 'Cast Lead,'" briefing, the Israel Fire and Combined Arms in Urban Terrain Conference, November 8–11, 2010.

[81] Amine Hoteit, "Harb Israeil ala Gaze wa Asaraha ala al-Istrategiya al-Aaskariya [The Gaza War and Its Impact on the Military Strategy]," Al-Jazeera Center for Studies, undated.

[82] Hoteit, "Harb Israeil ala Gaze wa Asaraha ala al-Istrategiya al-Aaskariya."

[83] Ibrahim al-Madhoune, "Israeil Hinama la Tantaser Touhzam wa al-Mouqawama hinama la Touhzam, Tantaser [Israel: Where It Does Not Win, It Is Defeated; The Resistance: Where It Is Not Defeated, It Wins]," The Palestinian Information Center, March 3, 2009.

ods. Most units were composed of only two or three fighters. Fighters had orders to conduct suicide operations if they found themselves exhausted, facing imminent death or capture, or in another unsustainable condition.[84]

Hamas actively sought to ambush IDF troops and to use IEDs and mines to limit IDF mobility. It also employed snipers, a tactic that Hezbollah did not often attempt in 2006.[85] Hamas also paid careful attention to the use of media, filming and disseminating footage of successful operations to boost morale among Hamas fighters and the Palestinian population and to create an environment of fear for the IDF and Israeli civilians.[86] Additionally, Hamas practiced disinformation through its "Ghost Martyrs," a cadre of members who leaked information that suicide bombers were hiding in some areas and waiting for IDF units to arrive, an activity designed to create confusion in the IDF and slow down its operations.[87]

Despite its preparations, Hamas was totally outmatched by the IDF. Yoram Cohen and Jeffrey White write that

> Hamas had planned to stand and fight, but the Qassam Brigades proved unequal to the task. Fairly early in the fighting, Hamas fighters began removing their uniforms and donning civilian clothing, further increasing the risk to the civilian population. Units in the field started to break down after a few days, then to disintegrate. Under the weight of IDF fire, . . . [al-Qassam Brigade] fighters hastily withdrew to the city for cover and concealment. Even in places where they were ordered to hold their positions they abandoned them, preferring to survive rather than to fight. None of their ground combat measures worked, and while this is not surprising, given the difficulties they faced, they cer-

---

[84] Hoteit, "Harb Israeil ala Gaze wa Asaraha ala al-Istrategiya al-Aaskariya."

[85] Hoteit, "Harb Israeil ala Gaze wa Asaraha ala al-Istrategiya al-Aaskariya."; IslamicNews. net, "Khottat al-Moqawama fi harb Gaza . . . Istrategiya Mahaliya."

[86] Al-Qassam, "Penetrating Arrow Security Operation," web page, undated.

[87] IslamicNews.net, "Khottat al-Moqawama fi harb Gaza . . . Istrategiya Mahaliya."

tainly failed to match the image Hamas tried so hard to present, of stalwart and proficient Islamic warriors.[88]

The End of Operation Cast Lead

On January 18, 2009, Israel accepted an Egyptian ceasefire proposal and began to withdraw IDF forces from Gaza. All IDF forces had left the Gaza Strip by January 21.

Overall casualty figures—particularly the numbers of Gaza civilians and Hamas fighters killed—are still hotly contested. A September 2009 UN report notes that,

> based on extensive field research, non-governmental organizations place the overall number of [Palestinian] persons killed between 1,387 and 1,417. The Gaza authorities report 1,444 fatal casualties. The Government of Israel provides a figure of 1,166.[89]

The 1,166 figure from the Government of Israel is from a March 2009 IDF accounting that notes that, "of the 1,166 names of Palestinian dead gathered by the IDF's Research Department, 709 have been identified as 'Hamas terror operatives.'"[90] The Palestinian Centre for Human Rights provides the following Palestinian casualty figures: "1,417 dead, including 926 civilians, 255 police officers, and 236 fighters."[91]

---

[88] Cohen and White, *Hamas in Combat*, p. 15.

[89] United Nations Human Rights Council, *Human Rights in Palestine and Other Occupied Arab Territories*, pp. 10–11.

[90] Yaakov Lappin, "IDF Releases Cast Lead Casualty Numbers," JPost.com, March 26, 2009.

[91] Palestinian Centre for Human Rights, "Confirmed Figures Reveal the True Extent of the Destruction Inflicted upon the Gaza Strip," press release, March 12, 2009. See also The Meir Amit Intelligence and Terrorism Information Center, *Fathi Hamad, the Hamas Administration's Interior Minister, Revealed that as Many as 700 Hamas Military-Security Operatives Were Killed During Operation Cast Lead*, November 3, 2010. In a November 1, 2010, interview with *Al-Hayat*, Fathi Hamad, Interior Minister of Hamas during Operation Cast Lead, seemed to confirm the higher Israeli estimates that "as many as 700 operatives from the military-security forces of Hamas and the other terrorist organizations were killed. Most

Israeli military casualties were "10 dead and 207 soldiers wounded and injured, of whom 19 severely."[92] Four of the deaths were caused by friendly fire.[93] Additionally, three Israeli civilians were killed, and seven were severely injured.[94]

Insights from Operation Cast Lead

The results of Operation Cast Lead were important for Israel for several reasons. First, the operation demonstrated that the IDF had corrected many of the deficiencies identified after the Second Lebanon War. Israel once again appeared to have a competent military, a fact that, at the strategic level, was central to meeting the Operation Cast Lead objective of restoring Israel's military deterrent. Additionally, the very nature of the operation showed that Israel could—and would—plan a campaign that included all elements of its power: air, ground, naval, intelligence, and the all-important signal of resolve of calling up reserves.[95]

Unresolved Problems in Determining Strategic "Ends"

The interface between key civilian and military leaders—Prime Minister Olmert, Deputy Prime Minister and Foreign Minister Livni, and Defense Minister Ehud Barak, on the one hand, and General Ashkenazi, on the other—seemed to have substantially improved since the Second Lebanon War. Nevertheless, as Cordesman notes, "Israel does not seem to have been properly prepared for the political dimen-

---

of them belonged to the Izz al-Din al-Qassam Brigades and some to the security forces, considered by Hamas as an integral part of the movement's military apparatus."

[92] Harel, "The Fire Delivery Concept in Operation 'Cast Lead.'"

[93] United Nations Human Rights Council, *Human Rights in Palestine and Other Occupied Arab Territories*, p. 11.

[94] Harel, "The Fire Delivery Concept in Operation 'Cast Lead.'"

[95] Author's discussions with IDF officers, Tel Aviv, February 10–12, 2009, and September 2–10, 2009; author's discussions with IDF officers, Washington, D.C., February 26, 2009, and April 1, 2009.

sions of war, or to have had any clear plan and cohesive leadership for achieving conflict termination."[96] There was friction among the top three civilian members about the conduct and termination of the war.[97] At a January 11, 2009, meeting, both Foreign Minister Livni and Defense Minister Barak reportedly advocated a quick termination of the operation. They were worried about incurring more IDF casualties and about causing more civilian casualties in Gaza City. In short, they believed that the gains made in strengthening Israel's military deterrent would be eroded—particularly in the diplomatic arena—by further action.[98] Prime Minister Olmert held tough, however, telling his full cabinet that "the pressure we are exerting [on Hamas] must not be reduced. Anyone who broadcasts weakness will earn the good will of the global community for 12 seconds, but will not change anything essential."[99] Prime Minister Olmert prevailed, and Operation Cast Lead continued until the unilateral Israeli ceasefire of January 18.

Reestablishing a credible military deterrent by demonstrating the competence and resolve of its ground forces was important to Israel, and it did seem to have an effect on other actors (i.e., Hezbollah) and countries (i.e., Iran and Syria) in the region. Hezbollah was quiet throughout Operation Cast Lead. Indeed, some within Israel argue that, strategically, the Second Lebanon War was much more successful than initially believed. Proponents of this view believe that the difficulties that Israel experienced in conducting ground operations in Lebanon prevented many Israelis from understanding how devastating the war had been for Hezbollah.[100] Indeed, when a small number of

---

[96] Cordesman, "The 'Gaza War,'" p. ii.

[97] Cordesman, "The 'Gaza War,'" pp. 28–29.

[98] Barak Ravid, "Olmert: Gaza War Won't End Until Rockets and Smuggling Stop," Haaretz.com, January 12, 2009.

[99] Ravid, "Olmert: Gaza War Won't End Until Rockets and Smuggling Stop."

[100] Author's discussions with IDF officers, Tel Aviv, February 10–12, 2009, and September 2–10, 2009; author's discussions with IDF officers, Washington, D.C., February 26, 2009, and April 1, 2009; author's discussions with IDF officers, Jerusalem, September 10, 2009. These arguments are similar to those advanced in Luttwak, "In Praise of Aerial Bombing," and Martin van Creveld, "Israel's War with Hezbollah Was Not a Failure," Forward.com, January 30, 2008.

rockets were fired into northern Israel during Operation Cast Lead, Hezbollah quickly denied responsibility. (Responsibility was eventually claimed by "a small radical Palestinian group" trying to "show solidarity with Hamas."[101])

Nevertheless, the Israeli parliamentary government's ability to state clear political objectives and war-termination criteria—the "ends" of strategy—was far from perfect in Operation Cast Lead. There was, and still is, considerable friction at the political and strategic levels, which persists in discussions in Israel about other regional challenges, such as what to do about Hezbollah in Lebanon. Despite the failure of IDF standoff attacks to force the Lebanese government to control Hezbollah in 2006, some in Israel maintain that "it is easier to fight another state than guerrillas" and favor a strategy that attempts to coerce the Lebanese government into dealing with Hezbollah (rather than having Israel attack Hezbollah directly).[102] The logic of this strategy, which would almost surely further weaken the Lebanese institutions that might eventually create an internal alternative to Hezbollah, is questionable. Additionally, this approach would create difficult diplomatic dilemmas for the United States and others that are trying to bolster the Government of Lebanon.

What is clear from Operation Cast Lead is that Israeli politicians provided sufficient guidance for the IDF to craft a military campaign with executable missions. Additionally, in contrast to the Second Lebanon War, political and military authorities in Tel Aviv did not meddle incessantly and micromanage operations.[103] An unanswered question remains, however: What about the next time?

Significant Improvement in the Military "Means"

The military "means" of achieving political objectives during Operation Cast Lead were vastly improved compared with those available

---

[101] Yaakov Katz, "Small Terror Group Said Behind Katyushas Fired at North," JPost.com, September 11, 2009.

[102] Author's discussions with IDF officers, Tel Aviv, September 2–10, 2009.

[103] Author's discussions with IDF officers, Tel Aviv, March 2–5, 2008, February 9–19, 2009, and September 2–10, 2009.

during the Second Lebanon War. The most-notable improvements occurred in the functioning of the General Staff, the regional command, and Israeli Army units and in interservice cooperation. Military commanders in the field followed a clearly laid out and phased plan that was generally well executed. Clearly, the IDF "was back."

Nevertheless, although experiences during the Second Lebanon War and Operation Cast Lead have given rise to broad agreement within the IDF that ground maneuver is a necessary component of any strategy to defeat hybrid opponents, there is rather public disagreement within the Israeli Army about what form a future ground action in Lebanon should take. At a September 2009 conference sponsored by the IDF Ground Forces Command and U.S. Joint Forces Command, Major General Avi Mizrachi, commander of IDF Ground Forces, stated that "a war cannot be won without moving forces on the ground. . . . Only a ground maneuver will end the conflict and win the war." Brigadier General Agay Yehezkeli, chief of the Armored Corps, expanded on General Mizrachi's view: "in a future conflict with Hizbullah in Lebanon the IDF would need to launch a quick ground operation, heavily depending on tanks, deep into Lebanese territory in order to curb the rocket attacks against the Israeli home front." Brigadier General Yossi Bahar, chief of paratroopers and infantry, voiced a decidedly different perspective, arguing that "a deep penetration of Lebanon was not needed immediately at the outset of the war" and that "several brigades would be capable of conquering southern Lebanon and taking control of the 165 villages south of the Litani River."[104]

There is also the point, made by Cordesman, that, although "the IDF did demonstrate that its ground forces have the military superiority or 'edge' in fighting asymmetric wars in the edges of a densely populated urban area . . . [,] the IDF did not pursue the ground war to any major tactical conclusion inside these areas."[105] Cordesman's assessment is correct, but it does not fully consider that the Israelis had no stomach for a more extensive operation, one that would have been

---

[104]Yaakov Katz, "Future Battlefield Will Be More Lethal," JPost.com, September 2, 2009. The author of this monograph was a speaker at the conference and witnessed this exchange.

[105]Cordesman, "The 'Gaza War,'" p. 68.

far more extensive and costly than Operation Cast Lead. Additionally, a broader operation would likely also have made Israel responsible for Gaza. These are central issues that go back to why Israel launched Operation Cast Lead in the first place. Indeed, Cordesman captures the Israeli view about restoring its deterrent through Operation Cast Lead when he quotes the following statement made by retired Major General Giora Elad: "This hasn't solved the problem. . . . But it has introduced a completely different cost calculation for Hamas."[106] Thus, the Israelis have a coping strategy that uses military force to create a deterrent against future actions from various actors when those actors' actions reach an unacceptable threshold. In the case of the Second Lebanon War, the soldier abductions triggered action. In Operation Cast Lead, the volume of rocket fire (particularly longer-range rockets), coupled with a need to demonstrate the competence of the IDF to reestablish Israel's regional deterrence, prompted action. Israel had no desire to get involved in protracted operations to fully defeat Hezbollah or Hamas. Indeed, it is important to remember that Israel withdrew from Lebanon in 2000 after a costly intervention and that it had turned Gaza over to the Palestinians in 2005. There was no desire in 2006 or 2009 to reoccupy these areas.

Operation Cast Lead brings into focus two key questions for Israel and others facing hybrid threats: First, how much and what type of military action is sufficient to create a deterrent that will successfully curb the actions (e.g., rocket and mortar launches into Israel) of such opponents as Hamas and Hezbollah? Second, how much blood and treasure is the state willing to invest to achieve this end? In the case of Operation Cast Lead, Israel placed clear limits on what it was willing to pay to affect Hamas' continued capabilities. Again, there was little appetite for trying to clear Hamas from Gaza City, and even less for having Israel reoccupy the Gaza Strip. The Israeli public is still averse to casualties, and a more-protracted ground campaign would almost certainly have resulted in more IDF casualties and in even greater international uproar.

---

[106]Cordesman, "The 'Gaza War,'" p. 68.

From the Israeli perspective, the course of action that combined standoff fires with "a small [ground] maneuver[,] which is at a low risk of casualties," was the best military choice in Gaza.[107] Furthermore, not causing further Palestinian casualties or more destruction in Gaza was important at the strategic level. It is reasonable to conclude that Israel has little appetite for occupying southern Lebanon (or more of Lebanon) in any future operation against Hezbollah.

Much of the IDF's military success during Operation Cast Lead can, as previously noted, be attributed to the return to an emphasis on "basics"—

> vastly improved . . . [and more-understandable] planning, training, and integration of air, ground, naval, and ISR capabilities. The ability to make these improvements, however, was fundamentally nested in an important [and fundamental] conceptual realization after the Second Lebanon War: Precision, stand-off fires are critical, but not sufficient, to . . . [defeating] hybrid warfare opponents. . . .[108]

The IDF realized that opponents like Hezbollah must be countered with a "joint, combined-arms approach that enables integrated fire and maneuver, particularly in complex terrain and in military operations [that occur 'amongst the people']." Although Hamas is qualitatively not as significant a challenge as Hezbollah, Operation Cast Lead showed that the IDF is now much more prepared for future hybrid warfare challenges than it was in 2006.[109]

---

[107] Author's discussion with an IDF officer, Tel Aviv, July 16, 2009.

[108] Johnson et al., *Preparing and Training for the Full Spectrum of Military Challenges*, p. 232. As one IAF officer noted, "There are things you cannot do from the air" (author's conversation with an IDF officer via telephone, January 17, 2011).

[109] Johnson et al., *Preparing and Training for the Full Spectrum of Military Challenges*, p. 232.

Military Lessons Learned from Operation Cast Lead

Operation Cast Lead provides lessons in several areas: military operations "amongst the people," joint, combined-arms fire and maneuver against hybrid opponents, and air-ground integration.

Military Operations "Amongst the People"

Operation Cast Lead is in many ways representative of the challenges sophisticated militaries face when conducting operations against an adversary that, like Hamas, is embedded "amongst the people," especially if, as was generally the case in Gaza, that population is supportive of the adversary. Additionally, as Cordesman writes,

> the fighting in Gaza is a case study in the fact that asymmetric warfare confronts any solider [sic] actually in combat with a constant stream of hard choices and exercises in situation ethics obscured by what Clausewitz called the "fog of war." In many cases, instant choices have to be made where all of the advances in intelligence and command and control do not allow those actually fighting to know the nature of the threat forces or the number of civilians at risk.

> At the same time, the very nature of asymmetric warfare often forces the weaker size [sic] to maximize this uncertainty by not wearing uniforms, mixing in civilian areas, and using collocated civilians—often women and children—to provide support. This is no more an act of cowardice than using the protection of a tank or aircraft, but it does mean that war is evolving in ways that often increase the risk of civilian casualties and put more and more strain on the capability of armed forces to limit those casualties.[110]

The IDF realized that executing Operation Cast Lead would necessarily require striking numerous targets in populated areas, particularly in Gaza City. It was also keenly aware that there would be intense external scrutiny of its operations and that public opinion, both inside

---

[110] Cordesman, "The 'Gaza War,'" p. 3.

Israel and internationally, would demand the careful use of force. Using force carefully required accurate information about target locations and, to avoid collateral damage and to minimize civilian casualties, precisely matching munitions to targets.[111] Additionally, the IDF conducted a comprehensive assessment of Gaza "by developing highly detailed maps that tracked Hamas movements, facilities, shelters and tunnels against civilian facilities, and the location of sensitive facilities like schools, hospitals, and religious sites."[112] Ironically, Hamas aided this effort by the very act of taking over Gaza from Fatah. Hamas became "visible" because it occupied government and other structures throughout Gaza, and the Israelis knew many of these locations.[113]

Israeli targeting efforts were also aided by extensive intelligence about Gaza that was based on years of surveillance by multiple means, including in-place agents. As previously noted, this intelligence effort was a cross-governmental activity between the IDF and other entities. Additionally, Israeli special operations forces were probably on the ground in Gaza conducting operations, liaising with informants, and directing air strikes.[114] One IDF officer stated that, during the operation, Hamas began shooting anyone on the street talking on a cell phone, believing they were relaying targeting information.[115]

The Israelis actively sought to warn Gaza residents of impending military action in an effort to encourage them to leave areas and buildings that were targeted by the IDF.[116] The UN Fact Finding Mission

---

[111] Author's discussions with IDF officers, Tel Aviv, February 10–12, 2009, and September 2–10, 2009; author's discussions with IDF officers, Washington, D.C., February 26, 2009. The United Nations Human Rights Council found that both Israel and Hamas committed war crimes and crimes against humanity before and during Operation Cast Lead.

[112] Cordesman, "The 'Gaza War,'" p. 17.

[113] Author's discussions with IDF officers, Tel Aviv, February 10–12, 2009, and September 2–10, 2009; author's discussions with IDF officers, Washington, D.C., February 26, 2009.

[114] United Nations Human Rights Council, *Human Rights in Palestine and Other Occupied Arab Territories*, p. 147.

[115] Author's discussions with IDF officers, Tel Aviv, February 9–19, 2009.

[116] Opall-Rome, "In Gaza, Both Sides Reveal New Gear"; Barbara Opall-Rome, "Maj. Gen. Ido Nehushtan: Commander, Israel Air & Space Force," DefenseNews.com, August 3, 2009.

on the Gaza conflict, citing Israeli documents, notes that the following measures were taken:

- The Israeli armed forces made 20,000 calls [to Gaza residents to warn them of impending Israeli attacks] on 27 December and 10,000 on 29 December 2008;
- 300,000 warning notes were dropped over the whole of the Gaza Strip on 28 December;
- 80,000 leaflets were dropped in Rafah on 29 December;
- In the context of the beginning of ground operations on 3 January, 300,000 leaflets were dropped in the entire Gaza Strip, especially in the northern and eastern parts;
- On 5 January, 300,000 leaflets were dropped in Gaza City, Khan Yunis and Rafah;
- In total some 165,000 telephone calls were made throughout the military operations;
- In total some 2,500,000 leaflets were dropped.

In addition to these measures, the Israeli Ministry of Foreign Affairs explains that the telephone calls were both direct calls and pre-recorded messages, that it made radio broadcasts, and that it developed a practice of dropping apparently light explosives on rooftops (referred to by some as "roof-knocking").[117]

However, the mission largely dismissed these warning measures as "not . . . the most effective possible in the circumstances" and doubted "that many were effective at all."[118] The UN mission also found that the IDF "had carried out intentional strikes against civilians."[119]

---

[117] United Nations Human Rights Council, *Human Rights in Palestine and Other Occupied Arab Territories*, p. 152. See also Harel, "81% of the Fire—Precision Guided Munitions," p. 20. Harel states that there were "almost 300,000 phone calls, 100,000 of which were answered."

[118] United Nations Human Rights Council, *Human Rights in Palestine and Other Occupied Arab Territories*, p. 161.

[119] United Nations Human Rights Council, *Human Rights in Palestine and Other Occupied Arab Territories*, p. 231.

It is difficult to determine how effective these warning efforts were. Israel wants to portray them as highly effective and careful, and Hamas has no incentive to verify Israeli efforts to minimize collateral damage and civilian casualties. Indeed, Hamas had an opposite aim in its efforts to shape international opinion during the conflict: It wanted to portray IDF attacks in as bad a light as possible.[120] The remarks of Colonel Richard Kemp, a former commander of British forces in Afghanistan, to the UN special session on the Goldstone Report are very insightful in this regard:

> During Operation Cast Lead, the Israeli Defence Forces did more to safeguard the rights of civilians in a combat zone than any other army in the history of warfare.
>
> Israel did so while facing an enemy that deliberately positioned its military capability behind the human shield of the civilian population. . . .
>
> Despite all of this, of course innocent civilians were killed. War is chaos and full of mistakes. There have been mistakes by the British, American and other forces in Afghanistan and in Iraq, many of which can be put down to human error. But mistakes are not war crimes.
>
> More than anything, the civilian casualties were a consequence of Hamas' way of fighting.[121]

In an April 1, 2011, opinion piece, Justice Richard Goldstone, head of the UN fact-finding mission that examined Operation Cast Lead, essentially reversed the conclusions presented in the mission's report, writing, "If I had known then what I know now, the Goldstone

---

[120]Author's discussions with IDF officers, Tel Aviv, February 10–12, 2009, and September 2–10, 2009; author's discussions with IDF officers, Washington, D.C., February 26, 2009, and April 1, 2009. See Cordesman, "The 'Gaza War,'" pp. 17–18.

[121]UNWatch.org, "U.K. Commander Tells UN Council: 'IDF Took More Precautions Than Any Military in History of Warfare,'" web page, October 16, 2009.

Report would have been a very different document. . . . civilians were not intentionally targeted [by the IDF] as a matter of policy."[122]

The IAF employed a wide variety of airborne sensors on both manned and unmanned platforms to continuously blanket Gaza with surveillance. These platforms provided visual and communications intelligence that was combined with human intelligence to identify targets during both the air and the air-ground phases of Operation Cast Lead.[123]

Precision guidance enabled the IAF to use smaller weapons to achieve the desired effects while also limiting collateral damage. The IAF extensively employed both Global Positioning System (GPS)- and laser-guided PGMs; as previously noted, nearly 81 percent of the approximately 5,400 bombs and missiles used in Gaza were precision guided, whereas only 36 percent were in the Second Lebanon War.[124] One of the weapons used was the new Israel Military Industries laser-guided hard-target penetration bomb, which is based on the 1,000-lb Mk-83 bomb; "bunker-buster" bombs were also employed.[125] Additionally, conventional Mk-82 500-lb bombs—modified by replacing their explosive filler with concrete or a mix of explosives and concrete—were used. Precision guidance made these weapons highly accurate, and the

---

[122]Richard Goldstone, "Reconsidering the Goldstone Report on Israel and War Crimes," WashingtonPost.com, April 1, 2011.

[123]Gil, "Operation 'Cast Lead'—Gaza."

[124]Harel, "81% of the Fire—Precision Guided Munitions," p. 20.

[125]David Eshel, "New Tactics Yield Solid Victory in Gaza," AviationWeek.com, March 11, 2009. See also Defense Industry Daily, "2007–08: Israel Looks to Replenish Bomb Stocks," DefenseIndustryDaily.com, August 7, 2007, which shows the breadth of the IAF weapon inventory in its details on a Department of Defense Security Cooperation Agency notification to Congress of a possible U.S. sale to Israel of

> 10,000 live MK-84 2,000-lb. bombs; 1,500 live MK-82 500-lb. bombs; 2,000 live BLU-109 2,000-lb. bombs with penetrator warheads; JDAM tail kits that add GPS/INS guidance to bombs (10,000); Paveway II laser-guidance kits for the 500-lb. MK-82 (2,500), the 1,000-lb. MK-83 (500), and the 2,000 pound MK-84 (1,000) bombs; 10,000 FMU-139 live fuze components; 10,000 FMU-152 live fuze components; and 50 GBU-28 Enhanced Paveway III 5,000-lb. "bunker buster" laser/GPS guided live bombs.

absence of or reduction in the amount of explosives limited collateral damage.[126]

The combination of exquisite intelligence, persistent surveillance, and precision munitions enabled very sophisticated attacks against targets. These attacks are perhaps best represented by the "roof-knocking" technique, which was employed when Hamas used human shields on top of buildings to deter IAF attacks. The following description of one such attack illustrates the IDF's detailed integration of ISR, attack, and C2:

- A multistory building was identified for destruction because it contained military materiel.
- A telephone call was placed to an occupant of the building, who was told to evacuate the building within ten minutes.
- An IAF UAV, attack helicopter, and fixed-wing fighter were assigned to the mission. They had the building in sight and were on a shared communications channel with both Southern Command and IAF headquarters in Tel Aviv. Throughout the attack, they discussed what was happening and what to do next.
- A short time after the telephone call, the occupants of the building appeared on the roof, becoming human shields intended to dissuade the Israeli attack.
- The IAF "knocked on the roof," shooting at a corner of the building with a missile from the attack helicopter in such a way as to not injure any of the individuals on the roof. At this point, the rooftop shields rapidly vacated the roof and exited the building.
- As they left the building they were counted. When the same number who had been on the roof were out the door and away from the building, the fighter was cleared to drop a PGM to destroy the building. A video of the attack showed secondary explosions, indicating the presence of explosives in the building.[127]

---

[126]Author's discussions with IDF officers, Tel Aviv, February 10–12, 2009, and September 2–10, 2009.

[127]Author's discussions with IDF officers, Tel Aviv, February 10–12, 2009, and September 2–10, 2009; author's discussions with IDF officers, Washington, D.C., February 26,

According to the IAF, this technique, used frequently during Operation Cast Lead, shows both some of the Israeli efforts taken to minimize civilian casualties and the thorough integration of all the participants in the attack.[128]

Joint Fire and Maneuver Against Hybrid Opponents

As was previously discussed, the IDF planned from the very beginning to use ground operations in Phase II of Operation Cast Lead, after the completion of the initial Phase I standoff fire attack. The IDF had learned a key lesson from its experience in Lebanon: Defeating hybrid opponents, particularly when they are operating in complex terrain, requires ground forces. A comment by made by IAF Major General (Reserve) Isaac Ben Israel to reporter Charles Levinson summarized the IDF's post–Second Lebanon War viewpoint: "What you're seeing today is a direct lesson of what went wrong in 2006. . . . In Lebanon we learned that if you want to stop these rocket launchers you need to send soldiers in and take the area and control it and this is what is being done now."[129]

The IDF ground forces that went into Gaza were well prepared to execute combined-arms fire and maneuver. In the years after the Second Lebanon War, they had been extensively trained and exercised in combat skills (including in joint exercises). IDF units also rehearsed their specific roles in Operation Cast Lead during the weeks preceding the ground attack, and IDF brigades were task organized with giving the units all the capabilities they needed (e.g., infantry, armor, engineers, air controllers, intelligence, artillery, attack helicopters, UAVs). Additionally, armored forces—tanks and heavily armored infantry carriers, adapted to survive against hybrid enemies by the addition of

---

2009. The IDF posted a number of videos on its website showing attacks on targets using human shields; see Israel Defense Forces, "Precision Airstrikes on Hamas Terror Targets 7 Jan. 2009," web page, January 7, 2009.

[128]Author's discussions with IDF officers, Tel Aviv, February 10–12, 2009, and September 2–10, 2009; author's discussions with IDF officers, Washington, D.C., February 26, 2009; author's conversation with an IDF officer via telephone, January 17, 2011.

[129]Charles Levinson, "Israel's Ground Assault Marks Shift in Strategy," Online.WSJ.com, January 5, 2009.

extra armor applied to vehicle bellies and elsewhere—played a key role in the operation. Used in conjunction with infantry, armored vehicles provided protected mobility (including for logistical forces and for medical evacuations occurring close to the front lines) and mobile firepower. In short, tanks and armored personnel carriers reduced risks and increased maneuver options.[130] The operation's focus was on brigade operations, and the brigade commanders were given the authority to execute within the plan. Given the terrain and the fleeting nature of the enemy, this approach was much more successful than the very centralized approach used in Lebanon.[131]

The IDF had learned in Lebanon that, in the absence of pressure from ground forces, its adversaries knew how to avoid detection and attack by overhead platforms. In Gaza, IDF ground maneuver "forced the enemy to react, to move, to expose himself. Taking them from amorphous in nature to shaped, which *is critical in an urban area*."[132] Thus, ground maneuver was critical in creating targets for ground and air fires. Fires were also important because they "paralyzed the enemy," thus fixing his position. This allowed IDF ground forces to close with Hamas fighters who were reluctant to expose themselves to attack from air or artillery.[133] The Israeli Navy, in addition to blockading Gaza, also provided fires and UAV support for ground forces.[134] These improvements in interservice integration were the result of significant corrective actions within the IDF—particularly between the IAF and the Israeli Army—after the Second Lebanon War.

---

[130] Author's discussions with IDF officers, Tel Aviv, February 10–12, 2009, and September 2–10, 2009; author's discussions with IDF officers, Washington, D.C., February 26, 2009, and April 1, 2009.

[131] Author's discussions with IDF officers, Tel Aviv, February 10–12, 2009, and September 2–10, 2009; author's discussions with IDF officers, Washington, D.C., February 26, 2009, and April 1, 2009; Gil, "Operation 'Cast Lead'—Gaza."

[132] Gil, "Operation 'Cast Lead'—Gaza."

[133] Author's discussions with IDF officers, Tel Aviv, February 10–12, 2009, and September 2–10, 2009; author's discussions with IDF officers, Washington, D.C., February 26, 2009, and April 1, 2009; Gil, "Operation 'Cast Lead'—Gaza."

[134] Cordesman, "The 'Gaza War,'" p. 18. In Gil, "Operation 'Cast Lead'—Gaza," we are told that the Paratroopers Brigade had a "Navy Cell" inside its brigade main headquarters.

Air-Ground Integration

One of the central IDF successes between the Second Lebanon War and Operation Cast Lead was its improvement of air-ground integration. The IAF controls almost all the aviation platforms in the IDF. (In the United States, the U.S. Army has its own helicopter fleet, including attack helicopters, and the U.S. Marine Corps has helicopters and fixed-wing aviation.) Thus, in the IDF, interservice cooperation between the IAF and the Israeli Army is the only way in which ground forces can get air support.

The majority of the improvements to air-ground integration after the Second Lebanon War were accomplished through organizational and procedural changes that used existing systems in a joint manner. As previously discussed, the IDF had removed air controllers from the brigades before the Second Lebanon War. The new system put in place before Operation Cast Lead introduced a significant IAF presence at the brigade, division, and regional command levels.

The key air-ground facilitation organization in the IDF is the Air-Ground Cooperation Unit, located in Tel Aviv. This unit is led by a colonel, and its mission is to prepare air controllers, help plan exercises with the Israeli Army, and take part in operational planning.[135]

The IDF's regional command headquarters generally include a full-time, active-duty lieutenant colonel from the IAF's cooperation and coordination unit. During wartime, the air staffing is significantly increased and usually includes augmentation by an IAF brigadier general or major general. Depending on the situation, this air cell can be staffed by as many as 40 officers (mostly reservists). During Operation Cast Lead, an active-duty brigadier general was the senior air officer in Southern Command headquarters.[136]

During Operation Cast Lead, there was also significant cooperation within Southern Command between artillery officers and air officers. One of the key functions of Southern Command was to manage

---

[135] Author's discussions with IDF officers, Tel Aviv, February 10–12, 2009; author's conversation with an IDF officer via telephone, January 17, 2011.

[136] Author's discussions with IDF officers, Tel Aviv, February 10–12, 2009, and September 2–10, 2009.

the target bank and match the best weapon to each target. The artillery officer in charge of this effort at Southern Command during Operation Cast Lead was the "Chief of Fire Essence." Sitting next to this officer was an IAF operations analyst who was an expert in IAF weapons. Having this expertise was critical for several reasons. First, some targets, such as specific buildings and tunnels, could be destroyed only by the larger bombs carried by fixed-wing aircraft. Second, after the initial artillery strikes at the beginning of the ground campaign, Israel decided—because of a desire to use only PGMs in order to limit collateral damage—to use the five 155-mm howitzer batteries available in Operation Cast Lead primarily to fire smoke projectiles in order to screen ground maneuver. (Field artillery also responded to counterfire missions from AN/TPQ-37 Firefinder radars.) Finally, because of the difficulty of identifying targets at any distance in urban terrain, artillery forward observers did not call for fire very often. Target identification and tracking were much more successful with overhead assets, particularly persistent UAV coverage.[137] The air control organizations in Israeli Army divisions and brigades are shown in Figure 3.3.

The brigade joint tactical air control (JTAC) section and the division joint forward air control (JFAC) section work within the fire/operations staff sections of those organizations. The senior air officer is a major in the brigade JTAC and is a lieutenant colonel in the division JFAC. The officers assigned to these organizations bring specific expertise about platforms; for example, an attack helicopter pilot advises the brigade commander on using attack helicopters. The principal difference between the brigade and division organizations, other than the rank of the senior air officer, is that the division usually includes a second fixed-wing officer and an IAF intelligence officer. There are no formal air control structures below the brigade level.[138]

During Operation Cast Lead, each maneuver brigade had an attack helicopter squadron and direct support from UAVs. Close air

---

[137] Author's discussions with IDF officers, Tel Aviv, February 10–12, 2009, and September 2–10, 2009.

[138] Author's discussions with IDF officers, Tel Aviv, February 10–12, 2009, and September 2–10, 2009.

Figure 3.3
IDF Air Control Organizations

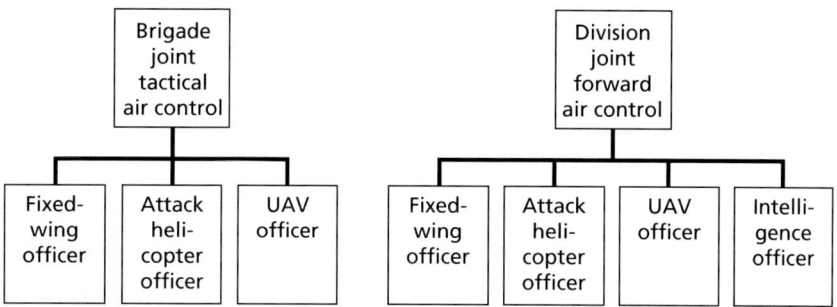

SOURCE: Provided to the author by the IAF in 2009.
RAND MG1085-3.3

support was almost always provided by attack helicopters that were responding to direction from the brigade JTAC or a maneuver battalion commander. In some cases, helicopters worked directly with maneuver companies and platoons. UAVs provided exceptional persistent situational awareness, and, as previously mentioned, their feeds were shared by different headquarters and platforms. Fixed-wing sorties were requested from a brigade JTAC through the division JFAC and were controlled by the IAF headquarters in Tel Aviv through the relevant brigade JTAC. There were three principal ways to request air sorties:

- through a preplanned request from the brigade to the division to the regional command to IAF headquarters in Tel Aviv
- through a real-time request from the brigade to the division to IAF headquarters in Tel Aviv (bypassing Southern Command)
- through direct-support attack helicopter requests from the relevant ground commander to the brigade JTAC.[139]

Supplementing the IDF's structural and procedural improvements were several other efforts, including deploying and using the follow-

---

[139] Author's discussions with IDF officers, Tel Aviv, February 10–12, 2009, and September 2–10, 2009.

ing: common language, maps with identical references (e.g., building numbers) and a coordinate system, dedicated radio nets, and improved target data bases and numbering systems.[140]

The vastly improved air-ground integration system employed by the IDF in Operation Cast Lead involved more than organizational and procedural changes.[141] At their core, the improvements were the result of air and ground officers working closely together to achieve a shared purpose, as noted by the IAF's commander during Operation Cast Lead, Major General Ido Nehushtan:

> The brigade commander was the one who had all the tools he needed and who commanded the operation, but professional considerations of how and under what conditions to employ the assets remained with the Air Force.
>
> Our pilots, air crews and UAV operators worked directly with ground commanders from the very early planning stages—each one in his own sector, to the point that they knew one another. They recognized each other's voices over the network and could smell each other's sweat. . . .
>
> [There was full] . . . partnership with the ground forces—from a moral and physical perspective—which required well-planned, well-rehearsed, truly joint operations based on a suit[e] of capabilities specifically sewn for their missions. Underlying all that is intimate cooperation between all relevant intelligence branches, which allowed commanders to constantly replenish their target banks during the course of the fighting.[142]

Interviews with Israeli Army and IAF officers confirmed that the spirit of cooperation constituted more than just general-officer talking points. One officer who had served in a brigade headquarters during Operation Cast Lead said that the IAF officers on his staff were "like

---

[140] Author's discussions with IDF officers, Tel Aviv, February 10–12, 2009, and September 2–10, 2009.

[141] Author's discussion with an IDF officer, Tel Aviv, September 8, 2009.

[142] Opall-Rome, "Maj. Gen. Ido Nehushtan."

waiters in a restaurant, frequently coming to ask him 'is everything OK?'" The cooperation was exceptional, and he knew the squadron commanders from the attack helicopter and UAV units supporting his brigade. Another officer emphasized the importance of the habitual association his brigade had with its supporting attack helicopter unit. They had trained together before Operation Cast Lead and knew and trusted each other. During the operation, air and ground officers sat at the same desk at headquarters and solved problems together.[143]

The IDF is still reviewing lessons from Operation Cast Lead. There is an ongoing dialogue between the Israeli Army and the IAF on the topic of positioning additional JTACs at the battalion level to further enhance air-ground integration. The IAF is concerned about such an increase in demand, however, because it is already stretching to meet the demands of the brigade air control structure. Ground officers also want to have greater access to fixed-wing aircraft and to bring the aircrafts' weapons closer to ground forces. Current IAF procedures require a 1-km separation between any fixed-wing strike and ground forces, a requirement resulting from a fratricide incident during a previous operation. The IAF position as of late 2009 is that attack helicopters are more appropriate for the close support role and that, given these assets and organic Israeli Army weapons, fixed-wing aircraft are unnecessary.[144]

There is also an ongoing discussion about control of air space. The Israeli Army currently controls all air space below 8,000 feet within its area of operations, which allows it to fire artillery and use its organic smaller UAVs. The Israeli Army uses fire-support-control measures (e.g., air-only corridors) to enable air operations. The IAF believes that this arrangement is cumbersome, particularly in the case of operations against fleeting targets.[145] The complexity of the air space C2 chal-

---

[143] Author's discussions with IDF officers, Tel Aviv, August 2–20, 2009.

[144] Author's discussions with IDF officers, Tel Aviv, August 2–20, 2009; author's email exchange with an IDF officer, November 6, 2009.

[145] Author's discussions with IDF officers, Tel Aviv, February 10–12, 2009, and September 2–10, 2009; author's conversation with an IDF officer via telephone, January 17, 2011.

lenges will likely increase as the Israeli Army continues to field additional UAVs.[146]

Finally, the Army is considering options to increase its organic fire-support capabilities. This is partly because it is not satisfied with IAF plans for providing air support to ground units. These new capabilities will likely include longer-range rocket and missile systems.[147]

Operation Cast Lead Was Not a Robust Hybrid Case

Operation Cast Lead, although it showed the remarkable progress the IDF had made since the Second Lebanon War, also identified several issues that could prove problematic for Israel—and other states who might fight hybrid opponents—in the future. These involve the nature of the adversary, the operating environment, resources, and future opponents. As more than one IDF officer told us, the Israelis are very aware that "Gaza is not Lebanon and Hamas is not Hezbollah."[148]

Hamas was not as difficult an adversary for the IDF as Hezbollah, partly because Hamas was not as well trained, organized, or equipped as Hezbollah. Hamas did not employ the quantity or quality of ATGMs and other weapons that Hezbollah did in 2006, presumably because such weapons were not available to it. This was particularly true in the case of military-grade rockets, of which Hamas had only a small number. Most of Hamas' rockets were the short-range Qassams that were, in the view of many IDF officers, more of a nuisance than a real threat.[149] As shown in Table 3.1, the locally produced Qassams have a relatively short range and carry small warheads. They are also not as accurate or reliable as the rockets employed by Hezbollah. The

---

[146] Author's email exchange with an IDF officer, December 10, 2010. This ground forces officer noted that "we [the Israeli Army] are in the beginning of developing with the IAF a new model for controlling the 'lower skies'—so it is going to be a complex procedure for us as well."

[147] Author's discussions with IDF officers, Tel Aviv, October 10, 2010.

[148] Author's discussions with IDF officers, Tel Aviv, February 10–12, 2009, and September 2–10, 2009; author's discussions with IDF officers, Jerusalem, September 10, 2009.

[149] Author's discussions with IDF officers, Tel Aviv, February 10–12, 2009, and September 2–10, 2009; author's discussions with IDF officers, Jerusalem, September 10, 2009.

main challenge posed to Israel by the Qassams was their psychological impact on the (relatively small) number of Israeli citizens within their range. Thus, they were more of a political weapon than a militarily significant threat. Additionally, before Operation Cast Lead, the Israeli government had worked to mitigate the effects of these weapons by adopting better civil defense measures and counseling its citizens. Finally, the Israeli government clearly focused on ensuring that better weapons did not make their way to Gaza, whether over the shores of the Mediterranean or through the tunnel network along the Egyptian border. It was able to prevent such movement to a considerable degree.[150] The impact of the overthrow of the government of Egyptian President Hosni Mubarak on the flow of weapons into Gaza remains to be seen. Obviously, the Israelis are concerned about the uncertain security situation in the aftermath of the "Arab Spring."

The IDF also had much better intelligence on Hamas than on Hezbollah. It was an area to which the Israeli intelligence apparatus had long devoted much attention and its full range of capabilities, from technical means to HUMINT. This better intelligence allowed Israel to conduct much better targeting against Hamas than it could against Hezbollah, and ground commanders entered Gaza with much better tactical intelligence. They also had active agents in Gaza that enabled targeting and battle damage assessment during the operation.[151] The operating environment in the air over Gaza was relatively benign and small.[152] UAVs and helicopters operated relatively freely with only some restrictions, and fixed-wing aircraft flew with impunity at altitudes above Hamas' capability to engage them. This gave them a decided advantage, as Cordesman notes:

---

[150] Author's discussions with IDF officers, Tel Aviv, February 10–12, 2009, and September 2–10, 2009; author's discussions with IDF officers, Jerusalem, September 10, 2009.

[151] Author's discussions with IDF officers, Tel Aviv, February 10–12, 2009, and September 2–10, 2009; author's discussions with IDF officers, Jerusalem, September 10, 2009.

[152] Author's conversation with an IDF officer via telephone, January 17, 2011. This officer noted that "small" can be a challenge, given the amount of air activity going on in a very tight air space.

The Israeli Air Force could also concentrate its assets over a small area, much of which was open or desert. Flight times were negligible and only limited assets have to be held in reserve to deal with the limited risk or intervention from outside states. The IAF's fixed wing aircraft could fly at high altitudes outside the line of sight, or where no one on the ground could know what or where a given aircraft could target. This allowed it to take full advantage of both advanced GPS and laser-guided munitions; and modern targeting avionics like synthetic aperture radar targeting pods, high resolution aerial imaging pods, and UAVs like the Shoval.[153]

Against a more competent and well-armed opponent, such freedom would likely not be possible. Indeed, the IAF lost a helicopter (most likely to MANPADS fire) in Lebanon during the Second Lebanon War.[154] If Hamas or Hezbollah were to obtain a significant MANPADS capability, that would certainly affect the IAF's ability to operate UAVs and helicopters. Furthermore, state actors that Israel might confront in the future (e.g., Iran and Syria) have air defense systems that are considerably more sophisticated than any the IAF would likely face in Lebanon or Gaza.

The terrain in Gaza was also more conducive to IDF operations than was the terrain in southern Lebanon. The flat, relatively open terrain in Gaza, coupled with the territory's small size, made a rapid ground advance to isolate Gaza City and sever the lines of communication with the Egyptian border a much simpler proposition than fighting through southern Lebanon to the Litani River, as was attempted in the latter stage of the Second Lebanon War. Additionally, the terrain in southern Lebanon is more complex, with its hills, channelizing terrain, and urban areas, and Hezbollah has more operational depth. Lebanon is a bigger area, roughly 45 km deep and 45 km wide at its

---

[153] Cordesman, "The 'Gaza War,'" p. 17.

[154] Hanan Greenberg, "24 Troops Killed in Day of Battles," YNetNews.com, last updated August 13, 2006. Hezbollah claimed it had shot down the helicopter with a Waad MANPADS missile, but the IDF attributed the loss to an ATGM. See also Harel and Issacharoff, *34 Days*, p. 233, where the authors state that the CH-53 helicopter was shot down by a SA-7 MANPADS.

widest point. Furthermore, the IDF had limited goals in Operation Cast Lead and chose not to try to occupy Gaza City. Finally, in Lebanon, unlike Gaza's Gaza City, there is not a single area that could have been isolated. In short, Lebanon would still pose formidable challenges to the IDF.

Operation Cast Lead was a relatively small-scale operation for the Israeli Army, involving only a handful of its best brigades, one division, and Southern Command. Thus, the IAF was able to resource the Israeli Army brigades in a way that would likely not be possible in a large-scale operation. For example, there are only four attack helicopter squadrons in the IAF, and one would run out of these squadrons very quickly if they were assigned in a direct support role to specific maneuver brigades, as they were in Operation Cast Lead.[155] Furthermore, the ability to deploy or assign JTACs and JFACs throughout the Israeli Army as fully as was accomplished in Operation Cast Lead would present significant challenges for the IDF. Resourcing battalion-level JTACs— which the Israeli Army wants—is obviously even more problematic.[156] This issue is magnified in the IDF because it is a conscript-based force. There is no professional, long-serving noncommissioned officer corps in the force. Male conscripts who do not become officers remain in the IDF for 36 months, and women serve for 24 months. Consequently, "growing" senior enlisted joint terminal attack controllers, like those in the U.S. Air Force, is likely not an option. Finally, given the fact that most JTAC and JFAC personnel are reserve officers, mobilized only for annual training periods and emergencies, maintaining proficiency would certainly present challenges.[157]

Operation Cast Lead also highlights two sources of continued Israeli vulnerability: rockets and missiles. Like Hezbollah during the Second Lebanon War, Hamas continued to launch rockets at Israel

---

[155] International Institute for Strategic Studies, *The Military Balance 2009*, p. 250.

[156] Author's discussions with IDF officers, Tel Aviv, February 10–12, 2009, and September 2–10, 2009.

[157] Author's discussions with IDF officers, Tel Aviv, March 2–5, 2008, February 10–12, 2009, and September 2–10, 2009; author's discussions with IDF officers, Jerusalem, September 10, 2009.

throughout Operation Cast Lead. Indeed, the very act of firing rockets in the face of Israeli military action was a way for Hamas and Hezbollah to show that they were not defeated and, thus, to both claim victory and put pressure on Israel to cease the war. As previously noted, some in Israeli consider Hamas' short-range rockets, given their inherent inaccuracy and small warheads, an almost tolerable nuisance; they are very difficult to find and attack, but their effects are fairly limited. Intermediate- and long-range rockets and missiles are, however, an entirely different matter. These unguided weapons and their conventional warheads are a significant problem. The introduction of guided rockets and missiles would make the problem even more dire.

Finally, Iran has become the central focus of Israeli security concerns. Many Israelis believe that Hezbollah and, to a lesser extent Hamas, are surrogates for Iran. Additionally, they realize that Iran, along with Syria, is the source of the high-end weapons and training that make Hezbollah a formidable adversary. Furthermore, the Israelis believe that their foremost security challenge is a nuclear-armed Iran. Many fear that an Iranian nuclear capability not only poses an existential threat in and of itself to the State of Israel but that it will also both embolden Hezbollah and Hamas to take more-aggressive action and destabilize the region in ways that are highly problematic for Israel. Indeed, more than one IDF officer told us that General Ashkenazi vowed when he became chief of staff that he would find a "solution to Iran" during his tenure.[158] As of this writing, it is unknown what form—a comprehensive missile defense system that shields Israel against the broad gamut of rockets and missiles that could be fired from Lebanon, Syria, Gaza, or Iran; offensive action against Iran; diplomacy; a combination of these approaches?—this solution might take.

---

[158] Author's discussions with IDF officers, Tel Aviv, February 10–12, 2009, and September 2–10, 2009; author's discussions with IDF officers, Jerusalem, September 10, 2009.

CHAPTER FOUR

The Relevance of the Israeli Experience for the U.S. Joint Force

Israel's security situation, like that of the United States, demands that it prepare for the full range of military operations—what the IDF calls the "rainbow of conflict." For the IDF, this spectrum includes low-intensity conflict (LIC), mainly focused on the West Bank; high-intensity conflict (HIC) against contiguous states, most notably Syria; and HIC against "states without borders" (specifically, Iran). What is clear from Lebanon and Gaza is that Israelis have expanded their view of what their defense forces must be ready to do. Before Lebanon, the Israeli Army focused almost exclusively on LIC, mostly in the form stopping the Second al-Aqsa Intifada and policing the West Bank and Gaza; the IAF was viewed as the national deterrent force for HIC, for targeted killing in LIC, and for retaliatory raids. The Second Lebanon War forced Israel to reevaluate this posture. It found in Hezbollah an opponent in "the middle" that created a qualitative challenge that requires forces with HIC skills employed at a lower scale (brigades) than in state-on-state conflict.

Hybrid opponents, and particularly ones like Hezbollah, can create conditions that are very similar to major combat operations. Taking a defended position from a well-armed and well-trained adversary with standoff fire capabilities, like Hezbollah, is HIC and requires sophisticated combined-arms fire and maneuver.

The operational methods of hybrid opponents—which operate in dispersed, small units and in complex terrain to avoid detection and attack from the air—also demand decentralized operations and mission command. Both the Second Lebanon War and Operation Cast Lead showed that combined-arms capabilities, ISR access, and air-ground integration have to be pushed to the brigade and lower levels to enable effective action against hybrid adversaries. These opponents are dispersed and are generally attempting to avoid decisive engagement by hitting hard and fading away. Thus, this enemy is fleeting, and micro-management and multiple command layers make the problem of find-ing, fixing, and capturing or killing these adversaries more difficult.

Toward a Capabilities-Based Understanding of Adversaries and Warfare

Writing about hybrid warfare after the Second Lebanon War became something of a cottage industry, and definitions of the term *hybrid war-fare* have proliferated. In an October 2009 article, Hoffman, attempt-ing to reintroduce precision into the discussion, defines a hybrid threat as "any adversary that simultaneously and adaptively employs a fused mix of conventional weapons, irregular tactics, terrorism and criminal behavior in the battle space to obtain their political objectives."[1] He exhorts us to avoid clinging to "oversimplified depictions of warfare in two distinct bins: conventional and irregular."[2]

---

[1]   Frank G. Hoffman, "Hybrid vs. Compound War: The Janus Choice—Defining Today's Multifaceted Conflict," *Armed Forces Journal*, October 2009.

[2]   Hoffman, "Hybrid vs. Compound War, the Janus Choice." Hoffman provides a useful appraisal of the state of professional discourse about hybrid warfare, citing the work of Gian Gentile, Bill Nemeth, Nathan Frier, Jack McCuen, Thomas Huber, T. X. Hammes, and Ralph Peters. He also discusses "compound wars" in this article, defining them as

> conflicts with regular and irregular components fighting simultaneously under unified direction. The complementary effects of compound warfare are generated by its ability to exploit the advantages of each kind of force and by the nature of the threat posed by each kind of force. The irregular force attacks weak areas and forces a conventional opponent to disperse his security forces. The conventional force generally induces the adversary to concentrate for defense or to achieve critical mass for decisive offensive operations.

This definition has been criticized by Huber, who writes that Hoffman "mischaracter-izes and trivializes" compound war (see Thomas M. Huber, "Huber Comments on Hybrid

Hoffman's description of the threat is important, principally because it should force a discussion of how "hybrid threats" should influence decisions about the military capabilities necessary to counter them. Again, however, one must be clear about what constitutes hybrid threat. As Gian Gentile writes, definitions have consequences, particularly for the U.S. Army:

> The danger of the terms "hybrid enemy" and "hybrid warfare" is that [they] are so vague, enabling people to define them any way that they want and then argue for a "transformed" army based on their particular interpretation of what constitutes a "hybrid" threat. . . . A greater danger is that using the concept "hybrid war" may "dumb down" clear and rigorous thinking about the future security environment.[3]

An examination of IDF performance in Lebanon and Gaza shows that what is important in designing the military means to contend with future threats is having a clear assessment of adversaries' military capabilities—both what they have and what they will likely not have. Thus, Hoffman's inclusion of terrorism and criminal behavior as necessary components of a hybrid adversary is too restrictive. Such a definition would exclude Viet Cong main force units during the Vietnam War as well as Yugoslavian and Soviet Union partisan forces during World War II. Instead, the more inclusive definition—that hybrid warfare is a "blend of the lethality of state conflict with the fanatical and protracted fervor of irregular war"—that Hoffman advanced in a 2007 book is more useful as a means to understand what capabilities are needed to prevail against such adversaries.[4] What is apparent from the Israeli experience is that there are opponents at three basic levels of

---

Warfare and Compound Warfare for DMH Faculty and Others Interested," post on the Combined Arms Center Blog, February 12, 2009).

[3] Gian P. Gentile, "The Imperative for an American General Purpose Army That Can Fight," *Orbis*, Vol. 53, No. 3, Summer 2009, p. 461. See also William F. Owen, "The War of New Words: Why Military History Trumps Buzzwords," *Armed Forces Journal*, November 2009, pp. 9, 35.

[4] Frank G. Hoffman, *Conflict in the 21st Century*, p. 28.

military capability and that each level places different demands on the military forces being designed to confront them.

A Capabilities-Based Model for Framing Current and Future Challenges

Figure 4.1 proposes a three-level model for understanding what military means are needed to deal with current and future adversaries.

Nonstate Irregular Adversaries

At the low end of the range of military operations, which U.S. doctrine largely clumps into the category "irregular warfare," are opponents like those the United States faced in Afghanistan and in Iraq during the height of the insurgencies in each country. There is an upper limit to the military capabilities these opponents can possess, especially in the absence of state sponsorship. Irregular adversaries generally consist of small groups of fighters with low levels of training who possess only short-range weapons (e.g., small arms, machine guns, RPGs, mortars, short-range rockets, and IEDs). Largely through acts of terror, they use their military means to induce instability in order to affect political conditions. Importantly, their weapons afford little standoff capability, and massing beyond squad-sized formations is unusual, given

Figure 4.1
Levels of Adversaries

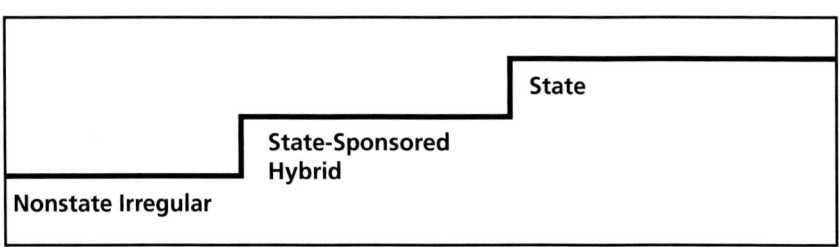

SOURCES: Johnson, *Military Capabilities for Hybrid War*; David E. Johnson and John Gordon IV, *Observations on Recent Trends in Armored Forces*, Santa Monica, Calif.: RAND Corporation, OP-287-A, 2010.
RAND *MG1085-4.1*

larger formations' vulnerability to air attack. In short, maneuver forces encounter irregular forces mostly in close combat (generally less than 1 km, and often less than 500 m) within the range of the latter's small arms, RPGs, and machine guns.

On occasion, irregular forces take advantage of the vulnerabilities of the opposing force to temporarily mass and attack, especially in the case of small outposts. This is what happened to the small outposts at Wanat and Keating in Afghanistan. After the attack, the Taliban forces disaggregated and slipped away. On the other hand, as in the Saddam Fedayeen forces that confronted the Coalition during Operation Iraqi Freedom, these irregular forces can be slaughtered by competent conventional forces when they attack. Figure 4.2 shows the general characteristics of irregular adversaries and lists recent representative cases.

The military capabilities needed to confront these opponents are those suited to coping with a long-term insurgency. The U.S. approach consists of COIN, stability operations, security assistance, and train-advise-assist missions. These are ground-centric, air-supported operations that require "boots on the ground." To avoid having a negative affect on external actors or on the population, whose support is so critical in COIN, fires (air and artillery) must be employed with precision to avoid collateral damage and civilian casualties. For force protection, air and artillery fires are generally used only as a last resort.

It is also important to note here the fundamental differences between the U.S. and Israeli approaches to dealing with irregular opponents. COIN, as the United States envisions it and as the classical authors prescribe, is usually about swinging the support of the populace in a favorable direction. In the occupied territories, however, the Israelis have no hope of doing this; instead, they generally "mow the grass" to suppress the enemy threat. Thus, one should be acutely aware that there are limitations on the lessons one can draw from Israeli experiences, particularly in the case of LIC and COIN.

Air Power in Irregular Warfare

Air power plays a critical role in irregular warfare, providing the overhead ISR, SIGINT, and strike capabilities critical to keeping the enemy from massing and for finding and killing high-value targets (HVTs).

Figure 4.2
Characteristics of Irregular Adversaries, with Examples

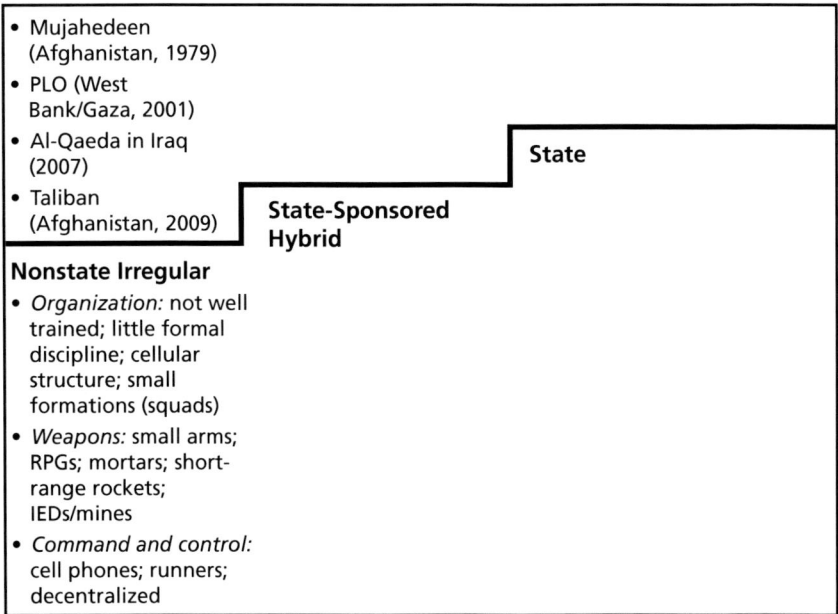

- Mujahedeen
  (Afghanistan, 1979)
- PLO (West
  Bank/Gaza, 2001)
- Al-Qaeda in Iraq
  (2007)
- Taliban
  (Afghanistan, 2009)

State

**State-Sponsored
Hybrid**

Nonstate Irregular
- *Organization:* not well
  trained; little formal
  discipline; cellular
  structure; small
  formations (squads)
- *Weapons:* small arms;
  RPGs; mortars; short-
  range rockets;
  IEDs/mines
- *Command and control:*
  cell phones; runners;
  decentralized

SOURCES: Johnson, *Military Capabilities for Hybrid War*; Johnson and Gordon,
Observations on Recent Trends in Armored Forces.
NOTE: PLO = Palestine Liberation Organization. RPG = rocket-propelled grenade.
RAND MG1085-4.2

Air Force air mobility capabilities are also key for the logistical support of the theater and for out-of-theater casualty evacuation. Army
and Marine Corps aviation provides fires and in-theater air mobility,
logistical support, and medical evacuation. In irregular warfare, to
prevent collateral damage and civilian casualties, most aerial fires will
be either ground-directed close air support or precision strikes against
HVTs. Furthermore, because the enemy's air defense capabilities are
generally limited to small arms, RPGs, and machine guns, air power
largely operates with impunity above 3,000 feet. U.S. forces also play

an important train, advise, and assist role by providing allies with their own air capabilities.[5]

Ground Power in Irregular Warfare

Ground forces in irregular warfare focus on establishing security for the population, developing HUMINT, and training, advising, and assisting indigenous forces so that they can eventually provide security for their own nation. Ground operations are conducted with the goal of clearing, holding, and building to enable the local host nation to supplant the insurgents. Friendly ground forces can be highly dispersed and may be located outside the range of organic fire support means (e.g., mortars, artillery). They also operate under restrictive rules of engagement that demand rigorous target identification. These realities of irregular warfare can increase the challenges of force protection, particularly if the enemy masses. Aerial fires—provided by both rotary- and fixed-wing aircraft—are an important means of providing responsive, timely, effective assistance.

Air-Ground Integration in Irregular Warfare

Irregular warfare is by definition highly ground centric, but airborne and space-based assets play a critical role as well. Furthermore, given the frequently decentralized and dispersed modes of maneuver in irregular warfare, air power is generally more effective when it is integrated at lower echelons (i.e., at the brigade combat team level). Table 4.1 depicts the roles of air power and ground power and the level of air-ground integration necessary in irregular warfare.

State-Sponsored Hybrid Adversaries

In the middle of the spectrum of operations are state-sponsored hybrid opponents. This is the type of adversary Israel faced in Hezbollah during the Second Lebanon War, that the Soviet Union eventually encoun-

---

[5] For discussion of air power in irregular warfare, see Robert C. Owen and Karl P. Mueller, *Airlift Capabilities for Future U.S. Counterinsurgency Operations*, Santa Monica, Calif.: RAND Corporation, MG-565-AF, 2007; and Alan J. Vick, Adam Grissom, William Rosenau, Beth Grill, and Karl P. Mueller, *Air Power in the New Counterinsurgency Era: The Strategic Importance of USAF Advisory and Assistance Missions*, Santa Monica, Calif.: RAND Corporation, MG-509-AF, 2006.

Table 4.1
Air and Ground Power in Irregular Warfare

	Irregular Warfare	State-Sponsored Hybrid Warfare	Deterrence/Major Combat Operations
The role of air power	• Overhead ISR and signals intelligence are crucial because the enemy does not mass. These capabilities are critical to finding and attacking high-value targets. • Air mobility is critical to supply and evacuation. • Air power is used mostly in ground-directed close air support with tight rules of engagement. It is key for force protection in extremis. • Air superiority is contested below 3,000 feet. Above 3,000 feet, air power is mainly invulnerable.	• Overhead ISR and signals intelligence are tightly linked with precision strike. • Air power is critical to attacking the enemy's deep strike assets and high-value targets. • Tight rules of engagement for centralized strikes and close air support are required. • Air power is used for the suppression of enemy standoff systems to support (complement) ground maneuver. • Air power complicates the enemy's ability to mass and be reinforced. • Air superiority may be contested below 20,000 feet.	• Air power is critical to deterrence achieved through global reach and strike capabilities. • Overhead ISR and signals intelligence are tightly linked with precision strike. • Strategic and operational air mobility and tankers are critical capabilities. • Air and space superiority may be contested at all levels. • Centralized control is critical. • Air power precludes large-scale ground maneuver by the enemy. • Air bases may be contested. • Operations may occur in a nuclear, biological, or chemical environment.

Table 4.1—Continued

	Irregular Warfare	State-Sponsored Hybrid Warfare	Deterrence/Major Combat Operations
The role of ground power	• Ground power is focused on establishing security, obtaining human intelligence, and training indigenous forces. • Maneuver is focused on clearing, holding, and building. • Dispersed operations increase the difficulty of force protection. • Tight rules of engagement demand rigorous target identification.	• Ground power is critical to forcing enemy reaction and to uncovering hidden assets. • Combined-arms operations are fundamental to success. • Ground power closes with enemy forces. • Ground power conducts decentralized operations against dispersed adversaries. • High-intensity MCO-like operations are possible at the brigade level and below. • Lines of communication may be vulnerable.	• Troop deployment is a key signal of national commitment. • Combined-arms operations are the key to success. • Ground maneuver forces an operational reaction from the enemy. • Ground power engages ground units that avoid air attacks and indirect fire. • Ground power is critical for exploiting operational opportunities and pursuing enemy forces. • Ground power deals with hybrid or irregular threats. • Ground power is critical to establishing post-MCO security and stability. • Basing and staging may be contested. • Operations may occur in a nuclear, biological, or chemical environment.
The level of air-ground integration	• Operations are ground-centric but highly dependent on air power. • C2ISR and joint tactical air controllers are best integrated at lower echelons for direct support.	• Balanced operation requires tighter coordination and extensive training and rehearsals. • Integration ensures high responsiveness to ground units and integration at levels below the theater.	• Air power control is highly centralized. • Air superiority is critical to ground maneuver. • Integrated suppression of enemy air defenses is key. • Supported-supporting relationships depend on the operation; the air or ground commander could lead. Leadership could change during an operation.

NOTE: C2ISR = command and control, intelligence, surveillance, and reconnaissance. MCO = major combat operations.

tered in the later years of its war in Afghanistan, and that Russia faced in Chechnya in the early 1990s. These adversaries pose a qualitatively different challenge than irregular opponents—a challenge that is similar to that posed by opponents in major combat operations, although it occurs on a smaller scale. The similarities between state-sponsored hybrid adversaries and opponents in major combat operations are due to the former's training, discipline, cohesion, organization, C2 capabilities, and weapons (e.g., ATGMs, MANPADS, intermediate- or long-range surface-to-surface rockets), which give them standoff fire capabilities. Irregular Palestinian forces operating during the Second al-Aqsa Intifada were generally engaged immediately in close combat at ranges of 500 meters or less. However, standoff weapons gave Hezbollah the capability to engage the IDF with mortars and ATGMs at extended ranges (as much as 5 km, in the case of AT-14 Kornet-E ATGMs). To successfully counter the Hezbollah threat, the IDF would have had to have used combined-arms fires to suppress the opponent's standoff weapons and thereby enable IDF infantry to maneuver into close combat ranges. Figure 4.3 shows the general characteristics of hybrid adversaries and lists recent representative recent cases.

To acquire the military capabilities that qualitatively separate them from irregular opponents, hybrid adversaries require state sponsors. State forces can become hybrid adversaries, a situation most likely to occur when a relatively weak state faces a threat from a more powerful neighbor. States supply hybrid adversaries with standoff weapons and training, as occurred in the cases of Hezbollah, which was supported by Iran and Syria, and the mujahedeen in Afghanistan in the 1980s, who received support from the United States, Pakistan, and others. In the slightly different case of Chechnya, the Chechens were trained during their service in the Soviet military before the dissolution of the Soviet Union, and, in their later conflicts with Russia, they mostly used weapons that had been stockpiled and left behind by the Soviets.

Compared with irregular adversaries, hybrid adversaries are generally capable of operating in larger formations with more coordination; this is because they are disciplined, have cohesion, have good C2

Figure 4.3
Characteristics of Hybrid Adversaries, with Examples

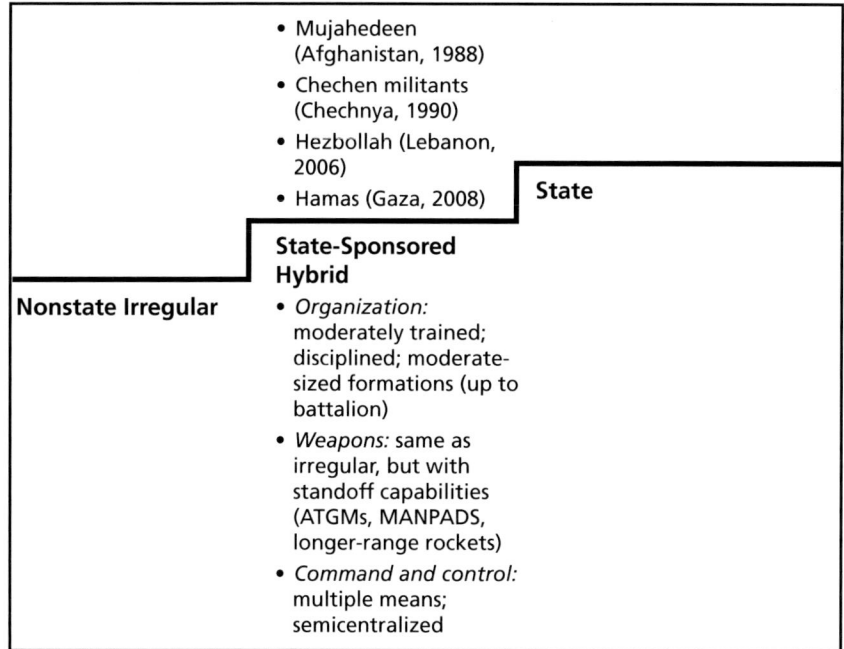

SOURCES: Johnson, *Military Capabilities for Hybrid War*; Johnson and Gordon, *Observations on Recent Trends in Armored Forces*.
RAND *MG1085-4.3*

capabilities, and understand how to avoid enemy strengths.[6] They are also willing and able to fight and stand their ground and can pose a coordinated military challenge across a large geographic area. This is perhaps the key difference between irregular and hybrid opponents: Hybrid opponents can make it difficult for you to take their territory; irregular opponents can generally only make you regret having done so.

---

[6] Cohesion is different from but obviously related to discipline. Cohesion may also be the key characteristic in determining whether certain types of irregular adversaries are more likely than others to become a viable hybrid threat if they receive state sponsorship. An insurgency with deep political roots and unified ideology, for example, might be more likely to pose a hybrid threat than might a warlord's militia.

Consequently, hybrid adversaries rely on dispersion, concealment, decentralized C2, and smaller formations than are typically employed by conventional forces. Hybrid adversaries make these choices hoping to employ asymmetric advantages to thwart their opponents. Indeed, weaker state actors might profitably employ these same approaches against states with more-advanced capabilities—particularly air and ISR—in order to blunt the latter's advantages. (One could argue, although it is beyond the scope of this study to do so, that the United States might most effectively assist potential partners—e.g., Georgia, Taiwan—that are facing significant state threats by providing them with hybrid capabilities rather than by helping them develop conventional small-state militaries.)

Perhaps the most important point in understanding how an irregular opponent makes the transition from irregular adversary to a hybrid adversary is this: *To become a hybrid adversary, an irregular force must have state support.* There appears to be a general—but erroneous—consensus that MANPADS are not only "easy to use" but also "readily available on the black market."[7] Although civilian and military aircraft have occasionally been brought down by MANPADS, these are properly thought of as *events* and are not necessarily indicative of adversary *capability*. In these cases, the terrorist or insurgent attackers obtained a limited number of MANPADS (sometimes only one)—something that should not be trivialized—but they did not possess a *capability*. To acquire MANPADS or other types of effective standoff weapons (e.g., ATGMs, rockets, shore-to-ship missiles) capability requires a state sponsor who provides the weapons, training, and supply chain needed to create and sustain the capability at a level that changes how the hybrid opponent's adversary can operate (e.g., limits the employment of helicopters, fixed-wing aircraft, and even unmanned aircraft systems, thus forcing changes in operational concepts that rely on helicopter-provided air support, troop movements, medical evacuation, and resupply). Thus, the casual view that "money talks" to the point that a nonstate actor can make itself as powerful as it can afford to pay

---

[7]   Federation of American Scientists, "Man-Portable Air Defense System (MANPADS) Proliferation: Understanding the Problem," web page, undated.

for is specious. (This is also true in the case of popular alarmist fears that nonstate actors will acquire weapons of mass destruction.) Acquiring these types of capabilities requires the help of states, and providing such capabilities leaves fingerprints.

Without state status, there is likely a limit to the kind of military capability that hybrid adversaries can attain. For example, long-range integrated air defenses, large ground formations, and air forces are likely far beyond their grasp, and this must be kept in mind while designing forces and strategies to counter hybrid opponents. Clearly, specific limitations depend on the size and sophistication of the actor. Nevertheless, most hybrid actors are not sufficiently big and well resourced to wield state-like military power. Indeed, opting for hybrid capabilities can place limits on military options; for example, focusing on defensive capabilities likely curbs the acquisition of offensive capabilities that would be needed to invade a neighbor.

The military capabilities needed to deal with hybrid opponents are similar to those deployed by Israel during Operation Cast Lead. They include standoff fires (principally delivered via air power), which are used to attack enemy installations, and key weapon systems and ground forces (supported by air, artillery, and mortars), which are used to root out and defeat enemy forces in complex terrain, particularly urban areas where adversaries fight "amongst the people." Table 4.2 depicts the roles of air power and ground power and the level of air-ground integration necessary in hybrid warfare.

Air Power in State-Sponsored Hybrid Warfare

Air power plays an important role in hybrid warfare. It provides overhead ISR and SIGINT, which, linked with responsive precision strike, enables attack of HVTs and "deep" targets. Air also suppresses enemy standoff systems to complement ground maneuver and denies the enemy the ability to mass or resupply forces. Furthermore, the likelihood that hybrid opponents will have MANPADS means that air superiority could be contested below 20,000 feet. The use of rotary-wing aircraft may be constrained in such an environment. Thus, fixed-wing aviation, operating above enemy air defenses, plays key roles in strike, ISR, and resupply of forward friendly forces. Finally, given the

Table 4.2
Air and Ground Power in State-Sponsored Hybrid Warfare

	Irregular Warfare	State-Sponsored Hybrid Warfare	Deterrence/Major Combat Operations
The role of air power	• Overhead ISR and signals intelligence are crucial because the enemy does not mass. These capabilities are critical to finding and attacking high-value targets. • Air mobility is critical to supply and evacuation. • Air power is used mostly in ground-directed close air support with tight rules of engagement. It is key for force protection in extremis. • Air superiority is contested below 3,000 feet. Above 3,000 feet, air power is mainly invulnerable.	• Overhead ISR and signals intelligence are tightly linked with precision strike. • Air power is critical to attacking the enemy's deep strike assets and high-value targets. • Tight rules of engagement for centralized strikes and close air support are required. • Air power is used for the suppression of enemy standoff systems to support (complement) ground maneuver. • Air power complicates the enemy's ability to mass and be reinforced. • Air superiority may be contested below 20,000 feet.	• Air power is critical to deterrence achieved through global reach and strike capabilities. • Overhead ISR and signals intelligence are tightly linked with precision strike. • Strategic and operational air mobility and tankers are critical capabilities. • Air and space superiority may be contested at all levels. • Centralized control is critical. • Air power precludes large-scale ground maneuver by the enemy. • Air bases may be contested. • Operations may occur in a nuclear, biological, or chemical environment.

Table 4.2—Continued

	Irregular Warfare	State-Sponsored Hybrid Warfare	Deterrence/Major Combat Operations
The role of ground power	• Ground power is focused on establishing security, obtaining human intelligence, and training indigenous forces. • Maneuver is focused on clearing, holding, and building. • Dispersed operations increase the difficulty of force protection. • Tight rules of engagement demand rigorous target identification.	• Ground power is critical to forcing enemy reaction and to uncovering hidden assets. • Combined-arms operations are fundamental to success. • Ground power closes with enemy forces. • Ground power conducts decentralized operations against dispersed adversaries. • High-intensity MCO-like operations are possible at the brigade level and below. • Lines of communication may be vulnerable.	• Troop deployment is a key signal of national commitment. • Combined-arms operations are the key to success. • Ground maneuver forces an operational reaction from the enemy. • Ground power engages ground units that avoid air attacks and indirect fire. • Ground power is critical for exploiting operational opportunities and pursuing enemy forces. • Ground power deals with hybrid or irregular threats. • Ground power is critical to establishing post-MCO security and stability. • Basing and staging may be contested. • Operations may occur in a nuclear, biological, or chemical environment.
The level of air-ground integration	• Operations are ground-centric but highly dependent on air power. • C2ISR and joint tactical air controllers are best integrated at lower echelons for direct support.	• Balanced operation requires tighter coordination and extensive training and rehearsals. • Integration ensures high responsiveness to ground units and integration at levels below the theater.	• Air power control is highly centralized. • Air superiority is critical to ground maneuver. • Integrated suppression of enemy air defenses is key. • Supported-supporting relationships depend on the operation; the air or ground commander could lead. Leadership could change during an operation.

high likelihood that hybrid opponents will operate among the populations and therefore will attempt to use actual or purported collateral damage as part of their information operations, rules of engagement will be restrictive and accurate target location information, precision and proportional strike, and battlefield forensics will be crucial.

Ground Power in State-Sponsored Hybrid Warfare

Ground forces are key to forcing hybrid adversaries to react and thereby make themselves vulnerable to detection and attack. Hybrid adversaries employ techniques that take advantage of complex terrain (e.g., urban, close) and prepared defensive positions that enable them to evade overhead observation. Combined-arms fire and maneuver that closes with hybrid defenses forces adversaries to come out from undercover and to fight or be overrun. At the brigade and lower echelons of ground formations, engagements against these hybrid forces resemble major combat operations in intensity (although not scale). Armored forces (i.e., tanks and armored infantry carriers) are important, given the lethality of enemy weapons (e.g., ATGMs, mortars, mines, IEDs). Finally, the nature of the hybrid battlefield creates conditions in which friendly lines of communication could be vulnerable to bypassed or hidden enemy forces.

Air-Ground Integration in State-Sponsored Hybrid Warfare

Because hybrid opponents possess effective standoff fire capabilities (e.g., ATGMs, MANPADS, mortars), defeating them requires joint, combined-arms fire and maneuver. Fire suppresses and fixes the enemy forces and supports ground forces as they close with the adversary. Fire also isolates the enemy, shutting off lines of supply and communication and limiting the adversary's ability to mass. Finally, maneuver forces enemy reaction. If enemy forces attempt to relocate to more-favorable terrain, they become visible and vulnerable to fire. If they remain in their positions and are suppressed, they can be defeated by ground maneuver. As stressed earlier, hybrid opponents demand integrated joint air-ground-ISR capabilities that are similar to those used against conventional adversaries but employed at a reduced scale.[8]

---

[8]   This monograph broadly defines *air-ground-ISR capabilities* as the constellation of air, ground, and space means used to find, fix, and capture or kill an adversary. To be truly effec-

Given the dynamics of the battlefield, the control of strike assets and maneuver decisionmaking has to be decentralized. Tight integration of the air-ground team in brigade and lower headquarters, extensive training, and rehearsals are fundamental for success. The Israelis learned this the hard way in the Second Lebanon War. Trying from headquarters in Tel Aviv to orchestrate and control highly decentralized engagements against Hezbollah fighters was ineffective.

Successful joint, combined-arms fire and maneuver requires a common purpose, a plan, and integrated C2 among all components. Consequently, joint, combined-arms training and operational rehearsals are necessary for success. Finally, fire and maneuver must be dynamically integrated, as shown in Figure 4.4.

Although there is likely an upper limit on the military capabilities of such hybrid opponents as Hezbollah, history shows that irregular opponents can move into the "middle" relatively easily. This was true in the case of the mujahedeen in Afghanistan in the 1980s. With training and good weapons, these irregular forces made the transition from irregular adversary to hybrid opponent and created difficulties for a Soviet military strategy heavily dependent on helicopters for fires and mobility.[9] In the current war in Afghanistan, all that the Taliban needs

---

tive, these capabilities must be integrated across services (and agencies), and the relevant information provided by the wide array of ISR resources must be readily available at the levels that need it.

[9] See Lester W. Grau and Michael A. Gress, eds., trans., *The Soviet Afghan-War: How a Superpower Fought and Lost*, Lawrence, Kan.: University of Kansas Press, 2002, p. 213. The authors note that

> the Mujahideen acquisition of the American-manufactured "Stinger" shoulder-fired air-defense missile gave the ability to hit an aircraft out to a distance of 4,800 meters and up to 2,000 meters in elevation. The Soviet command had to severely limit the use of helicopters, especially during daylight.

The acquisition of improved weapons by their adversary was only part of the problem the Soviets faced. Although Western weapons created serious tactical and operational challenges for the Soviets, the fundamental strategic reasons that they chose to withdraw from Afghanistan had to do with their inability to isolate the insurgency from its sanctuary, lack of support from the populace, and unwillingness to sustain a Soviet and Afghan force large enough to control the country and defeat the insurgency.

Figure 4.4
Joint, Combined-Arms Fire and Maneuver

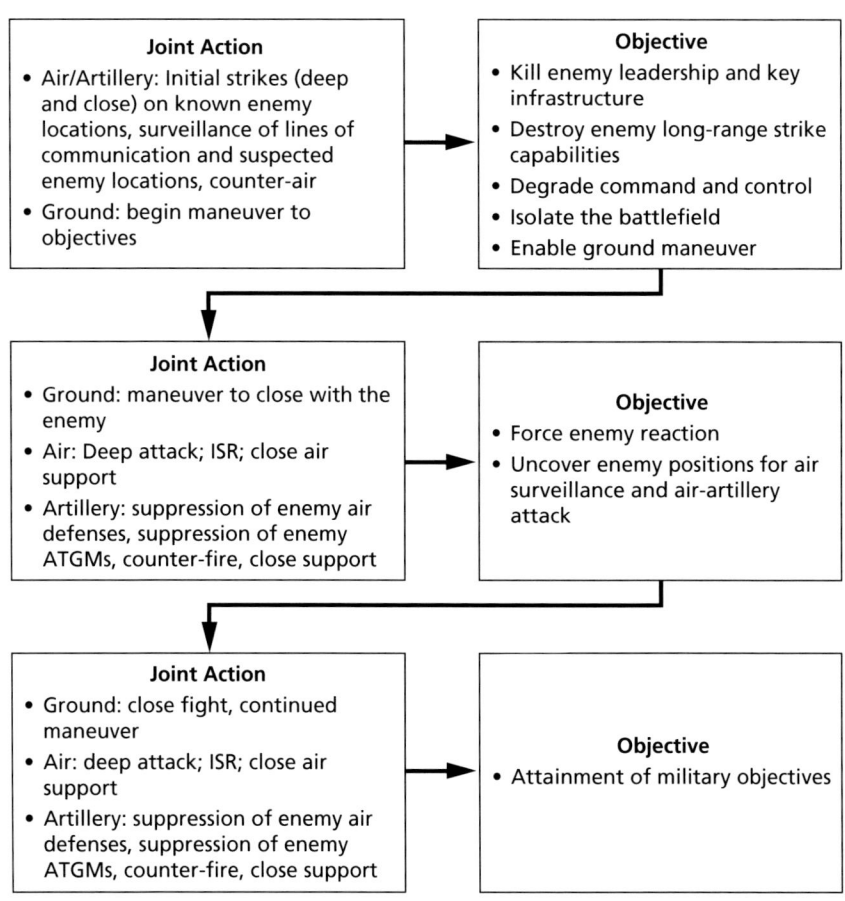

Joint Action
- Air/Artillery: Initial strikes (deep and close) on known enemy locations, surveillance of lines of communication and suspected enemy locations, counter-air
- Ground: begin maneuver to objectives

Objective
- Kill enemy leadership and key infrastructure
- Destroy enemy long-range strike capabilities
- Degrade command and control
- Isolate the battlefield
- Enable ground maneuver

Joint Action
- Ground: maneuver to close with the enemy
- Air: Deep attack; ISR; close air support
- Artillery: suppression of enemy air defenses, suppression of enemy ATGMs, counter-fire, close support

Objective
- Force enemy reaction
- Uncover enemy positions for air surveillance and air-artillery attack

Joint Action
- Ground: close fight, continued maneuver
- Air: deep attack; ISR; close air support
- Artillery: suppression of enemy air defenses, suppression of enemy ATGMs, counter-fire, close support

Objective
- Attainment of military objectives

RAND *MG1085-4.4*

to become a more lethal adversary is a state supporter that will supply it with effective standoff weapons and training in their use.[10]

---

[10] For a brief discussion of the Soviet Union in Afghanistan, see David E. Johnson, Adam Grissom, and Olga Oliker, *In the Middle of the Fight: An Assessment of Medium-Armored Forces in Past Military Operations*, Santa Monica, Calif.: RAND Corporation, MG-709-A, 2008. See also David E. Johnson, "Iran's Counter-Strike," *Providence Journal*, February 25, 2010.

State Adversaries

The capabilities of state adversaries can vary considerably, ranging from those with relatively low levels of capability (e.g., Georgia) to those with a wide array of capabilities, including nuclear weapons (e.g., the United States, Russia, China). Figure 4.5 shows the general characteristics of state adversaries and lists recent representative recent cases.

When Operation Iraqi Freedom began, and during its early years, the U.S. military was largely designed to deter and defeat state adversaries. Indeed, the conventional prowess of such top-tier conventional militaries as that of the United States has driven some states (e.g., North Korea, Iran) to pursue nuclear weapons as a hedge: They realize that they will never have the conventional ability needed to successfully counter the United States. Furthermore, state actors can always choose

Figure 4.5
Characteristics of State Actors, with Examples

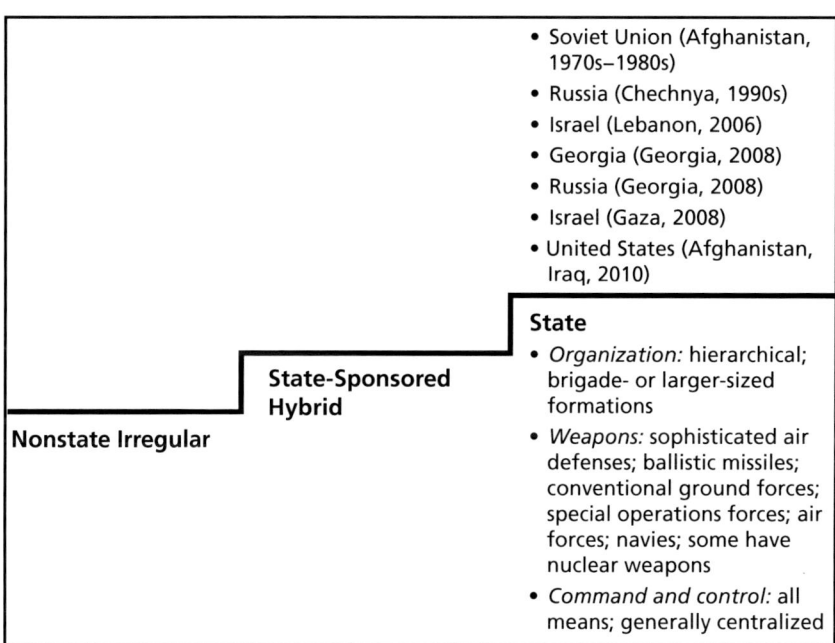

SOURCES: Johnson, *Military Capabilities for Hybrid War*; Johnson and Gordon, *Observations on Recent Trends in Armored Forces.*
RAND *MG1085-4.5*

to employ lesser capabilities within their arsenals if it makes sense operationally. Thus, in an example of what some call *compound warfare*, Russia employed high-end conventional capabilities (i.e., armored formations and bombers) against Georgia but also used cyber attacks and "irregular Chechen units, including the notorious Vostock Battalion."[11] This was a choice, but one must also recognize that what crushed the Georgian military was overwhelming Russian force, which included Russian air, naval, ground, and special forces units.[12] Quite simply, the Russians ratcheted up their efforts until they prevailed.

In designing capabilities to deter or fight state actors, one must appreciate that these actors have a wide array of options (including, in the case of Russia and others, nuclear weapons) that they can *choose to use*.

The roles of air power and ground power—and their integration—are influenced by state adversaries in the ways shown in Table 4.3.

Air Power in Deterrence/Major Combat Operations

Air power has the unique capability to operate across a theater (and, in the case of the United States, globally) to deter and strike using both conventional and nuclear means. The environment in which friendly air power operates against a state actor depends to a great degree on the quality of the specific state adversary. In some instances, such as the war in Georgia, the state's ability to control its airspace can be overwhelmed rather rapidly. In other cases, such as against a more formidable competitor, air and space superiority will be contested at all levels. Furthermore, given the range and precision of existing and future cruise missiles, basing will be contested—something that Israel began to understand during the Second Lebanon War. Additionally, the likelihood that operations may have to be conducted in nuclear, biological, or chemical environments must be a fundamental consideration.

In major combat operations, air power will likely be a constrained resource. It will also be in high demand. Therefore, centralized con-

---

[11] Hoffman, "Hybrid vs. Compound War—the Janus Choice."

[12] Roger N. McDermott, "Russia's Conventional Armed Forces and the Georgian War," *Parameters*, Spring 2009, pp. 65–80.

Table 4.3
Air and Ground Power in Deterrence/Major Combat Operations

	Irregular Warfare	State-Sponsored Hybrid Warfare	Deterrence/Major Combat Operations
The role of air power	• Overhead ISR and signals intelligence are crucial because the enemy does not mass. These capabilities are critical to finding and attacking high-value targets. • Air mobility is critical to supply and evacuation. • Air power is used mostly in ground-directed close air support with tight rules of engagement. It is key for force protection in extremis. • Air superiority is contested below 3,000 feet. Above 3,000 feet, air power is mainly invulnerable.	• Overhead ISR and signals intelligence are tightly linked with precision strike. • Air power is critical to attacking the enemy's deep strike assets and high-value targets. • Tight rules of engagement for centralized strikes and close air support are required. • Air power is used for the suppression of enemy standoff systems to support (complement) ground maneuver. • Air power complicates the enemy's ability to mass and be reinforced. • Air superiority may be contested below 20,000 feet.	• Air power is critical to deterrence achieved through global reach and strike capabilities. • Overhead ISR and signals intelligence are tightly linked with precision strike. • Strategic and operational air mobility and tankers are critical capabilities. • Air and space superiority may be contested at all levels. • Centralized control is critical. • Air power precludes large-scale ground maneuver by the enemy. • Air bases may be contested. • Operations may occur in a nuclear, biological, or chemical environment.

Table 4.3—Continued

	Irregular Warfare	State-Sponsored Hybrid Warfare	Deterrence/Major Combat Operations
The role of ground power	• Ground power is focused on establishing security, obtaining human intelligence, and training indigenous forces. • Maneuver is focused on clearing, holding, and building. • Dispersed operations increase the difficulty of force protection. • Tight rules of engagement demand rigorous target identification.	• Ground power is critical to forcing enemy reaction and to uncovering hidden assets. • Combined-arms operations are fundamental to success. • Ground power closes with enemy forces. • Ground power conducts decentralized operations against dispersed adversaries. • High-intensity MCO-like operations are possible at the brigade level and below. • Lines of communication may be vulnerable.	• Troop deployment is a key signal of national commitment. • Combined-arms operations are the key to success. • Ground maneuver forces an operational reaction from the enemy. • Ground power engages ground units that avoid air attacks and indirect fire. • Ground power is critical for exploiting operational opportunities and pursuing enemy forces. • Ground power deals with hybrid or irregular threats. • Ground power is critical to establishing post–MCO security and stability. • Basing and staging may be contested. • Operations may occur in a nuclear, biological, or chemical environment.
The level of air-ground integration	• Operations are ground-centric but highly dependent on air power. • C2ISR and joint tactical air controllers are best integrated at lower echelons for direct support.	• Balanced operation requires tighter coordination and extensive training and rehearsals. • Integration ensures high responsiveness to ground units and integration at levels below the theater.	• Air power control is highly centralized. • Air superiority is critical to ground maneuver. • Integrated suppression of enemy air defenses is key. • Supported-supporting relationships depend on the operation; the air or ground commander could lead. Leadership could change during an operation.

trol is essential. Air power can also be decisive, hitting strategic targets and denying the enemy the ability to maneuver or mass. It also plays a critical role in precision strike, close air support, interdiction, logistical support, strategic and operational air mobility, and ISR.[13]

Ground Power in Deterrence/Major Combat Operations

The employment of ground forces is, first and foremost, a key signal of resolve and commitment. As in the case of air power, basing and staging related to ground power will likely be contested by enemy anti-access forces that may employ nuclear, biological, or chemical weapons.

Ground force maneuver can be fundamental to forcing enemy reaction, thus making adversary forces vulnerable to destruction by fires or close combat. Combined-arms fire and maneuver are also key in exploiting operational opportunities and in pursuing enemy forces that survive the close fight or are bypassed. Finally, as evidenced by operations in Iraq and Afghanistan, ground forces play a strategic role in achieving desired end states by establishing the post–major combat operations security environment. They must also contend with irregular and hybrid threats that can arise when the state adversary collapses.

Air-Ground Integration in Deterrence/Major Combat Operations

As previously noted, air power is highly centralized in major combat operations. Furthermore, if the situation requires it, the air component may be the supported force. (This was the case in Operation Iraqi Freedom in the western regions of Iraq.) Finally, control of the air is an extremely desirable precondition for successful ground operations. An entry in Field Marshall Erwin Rommel's diary during the 1944 battle in Normandy bears repeating:

> During the day, practically our entire traffic . . . is pinned down with powerful fighter-bomber and bomber formations, with the result that the movement of our troops on the battlefield is almost completely paralysed, while the enemy can manœuvre freely. Every traffic defile in the rear areas is under continual attack and

---

[13] For a broader discussion of major combat operations, see Johnson, *Learning Large Lessons.*

it is very difficult to get essential supplies of ammunition and petrol up to the troops.[14]

Finally, given that the airspace can be highly contested, collaborative joint suppression of enemy air defenses (SEAD) is fundamental to air operations. Effective SEAD requires highly integrated cross-service efforts, and ground systems have an important role to play in these efforts.[15]

Being Effective Against One Type of Adversary Does Not Necessarily Prepare a Force for the Full Range of Operations

The Second Lebanon War taught the IDF that, although its capabilities were highly effective against state adversaries, they did not translate into immediate success in prosecuting wars against hybrid adversaries. Similarly, Operation Iraqi Freedom taught the U.S. military that, although its capabilities were, like Israel's, highly effective against state adversaries, they did not translate into immediate success in prosecuting wars against irregular opponents. Avi Kober writes about the irony of this situation in his assessment of Israel's performance in the Second Lebanon War:

> It is more logical to assume that against a regular army of a state, such as the Syrian Army, where the IDF could make far more effective use of its technological edge, a battlefield decision was attainable. Defeating a conventional opponent is easier for the IDF than winning a war against a diffused enemy like Hizballah, just like it was easier for the US-led coalition to defeat Saddam

---

[14] B. H. Liddell Hart, *The Rommel Papers*, New York: Harcourt, Brace, 1953, p. 477.

[15] For more on air-ground integration, see Jody Jacobs, David E. Johnson, Katherine Comanor, Lewis Jamison, Leland Joe, and David Vaughan, *Enhancing Fires and Maneuver Capability Through Greater Air-Ground Joint Interdependence*, Santa Monica, Calif.: RAND Corporation, MG-793-AF, 2009.

Hussein's armed forces than to put an end to the LIC that followed the high-intensity conflict in Iraq.[16]

Kober continues with an assessment of the implication of the Second Lebanon War for states that were overly enthusiastic about the RMA:

> And indeed, the Israeli case is representative of both Western democratic and high-technology countries waging asymmetrical wars. It is a warning sign against the over-reliance on technology in general and on airpower or network-centric warfare in particular, or the illusion that thanks to technology such countries can rely on "small but smart" militaries, and that technology minimizes fatalities, eliminates friction, decreases the dependence on logistics, breaks the enemy's will and can achieve quick victory by itself. RMA conceptions may be elegant and sophisticated, but they cannot replace simple military notions that have been held by military thinkers for centuries, such as the identification of and operation against centers of gravity—not just creating "effects"; the role played by ground forces in battlefield success; the importance of inflicting physical damage on the enemy—not just "burning its consciousness"; and the fact that the enemy does not abide by the rules one wishes to dictate.[17]

Indeed, the IDF is concerned that Syria may be adapting, along Hezbollah-like lines, large parts of its ground forces into hybrid formations with the goal of presenting an asymmetric challenge to the IDF's conventional air and ground prowess.[18]

As previously discussed, transitioning from one level of capability to another requires a marked increase in an adversary's military capabilities. Furthermore, moving from irregular to hybrid generally requires the sponsorship of a state actor. Thus, an important strategic consideration is how to deter a state actor from providing capabilities to irregular actors, thus preventing the latter from becoming hybrid

---

[16] Kober, "The Israel Defense Forces in the Second Lebanon War," p. 38.

[17] Kober, "The Israel Defense Forces in the Second Lebanon War," p. 38.

[18] Author's discussions with IDF officers, Latrun, Israel, September 2–3, 2009.

actors. The most difficult transition is the one from hybrid to state, and, as previously noted, there is wide variance in the capabilities of state actors.[19]

Militaries that become highly focused on irregular warfare tend to become very competent at related operations. However, in doing so, they can lose their ability to execute combined-arms fire and maneuver at the levels of proficiency and scale required to operate effectively against such hybrid opponents as Hezbollah. As Israel learned in fighting the Second al-Aqsa Intifada, and as the U.S. military has learned in Iraq and Afghanistan, the imperative to conduct protracted irregular warfare operations with limited ground forces inevitably leads to the need to make choices about training focus and combat preparation. These choices can affect the readiness of the force to deal with hybrid and state threats.

Figure 4.6 and Table 4.4 aggregate the preceding discussion of the three levels of adversary capability and the associated ground, air, and air-ground considerations.

What Lessons Should the U.S. Joint Force Draw from the IDF's Experiences?

The Israeli experience in the Second Lebanon War shows the challenges that hybrid threats, such as Hezbollah, can pose. One of the key insights to take from the Israeli experience is that the IDF was very good at what it was doing before the war but that the Israeli Army was extremely stretched to meet the everyday demands it faced in the West Bank and Gaza. Furthermore, Israel's strategic assessment prior to 2006 was that conducting low-intensity, asymmetric operations—such as those in the West Bank and Gaza—would remain the main role of the Israeli Army in the future. The IAF would deter state adversaries, and Israeli Army reserves would have time to mobilize and train in

---

[19] See David E. Johnson, Karl P. Mueller, William H. Taft, *Conventional Coercion Across the Spectrum of Operations: The Utility of U.S. Military Forces in the Emerging Security Environment*, Santa Monica, Calif.: RAND Corporation, MR-1494-A, 2002.

Figure 4.6
Levels of Adversaries and Their Associated Military Capabilities

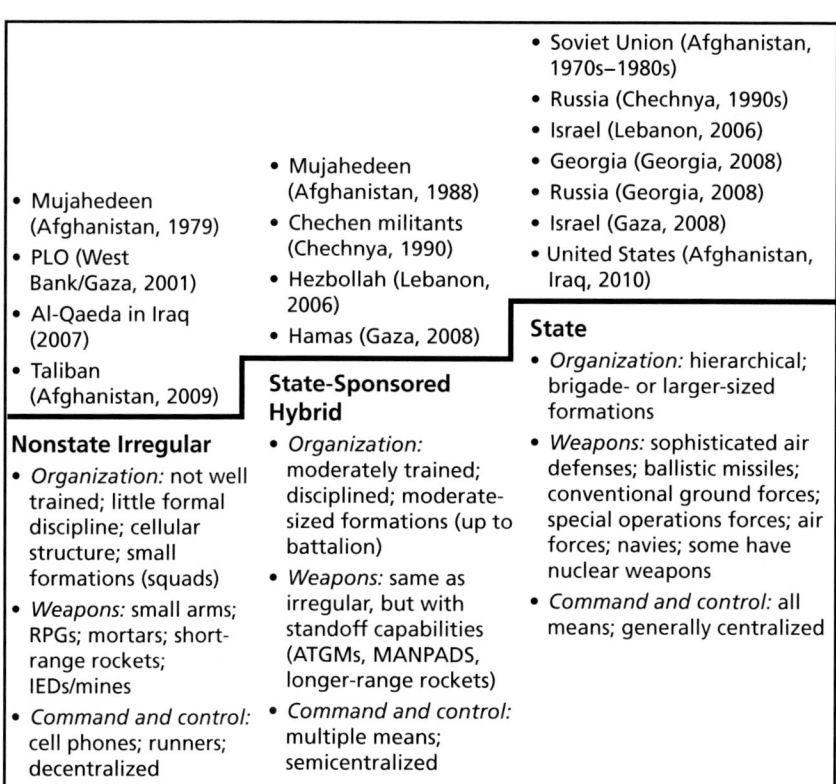

- Mujahedeen (Afghanistan, 1979)
- PLO (West Bank/Gaza, 2001)
- Al-Qaeda in Iraq (2007)
- Taliban (Afghanistan, 2009)

Nonstate Irregular
- *Organization:* not well trained; little formal discipline; cellular structure; small formations (squads)
- *Weapons:* small arms; RPGs; mortars; short-range rockets; IEDs/mines
- *Command and control:* cell phones; runners; decentralized

- Mujahedeen (Afghanistan, 1988)
- Chechen militants (Chechnya, 1990)
- Hezbollah (Lebanon, 2006)
- Hamas (Gaza, 2008)

State-Sponsored Hybrid
- *Organization:* moderately trained; disciplined; moderate-sized formations (up to battalion)
- *Weapons:* same as irregular, but with standoff capabilities (ATGMs, MANPADS, longer-range rockets)
- *Command and control:* multiple means; semicentralized

- Soviet Union (Afghanistan, 1970s–1980s)
- Russia (Chechnya, 1990s)
- Israel (Lebanon, 2006)
- Georgia (Georgia, 2008)
- Russia (Georgia, 2008)
- Israel (Gaza, 2008)
- United States (Afghanistan, Iraq, 2010)

State
- *Organization:* hierarchical; brigade- or larger-sized formations
- *Weapons:* sophisticated air defenses; ballistic missiles; conventional ground forces; special operations forces; air forces; navies; some have nuclear weapons
- *Command and control:* all means; generally centralized

SOURCES: Johnson, *Military Capabilities for Hybrid War*; Johnson and Gordon, *Observations on Recent Trends in Armored Forces*.
RAND MG1085-4.6

the unlikely event of a war with a bordering state—a possibility made even less likely by the ongoing U.S. presence in Iraq. Budgets were cut, organizations were optimized for this projected future, and the Israeli Army staffs and processes for integrating air and ground operations were removed (especially from brigades). The IDF also adopted operational concepts that took advantage of lessons from U.S. operations in Kosovo, Afghanistan, and Iraq. The IDF emulated the high-technology approaches from these wars because they promised rapid

Table 4.4
Air and Ground Power Across the Levels of Adversary Capabilities

	Irregular Warfare	State-Sponsored Hybrid Warfare	Deterrence/Major Combat Operations
The role of air power	• Overhead ISR and signals intelligence are crucial because the enemy does not mass. These capabilities are critical to finding and attacking high-value targets. • Air mobility is critical to supply and evacuation. • Air power is used mostly in ground-directed close air support with tight rules of engagement. It is key for force protection in extremis. • Air superiority is contested below 3,000 feet. Above 3,000 feet, air power is mainly invulnerable.	• Overhead ISR and signals intelligence are tightly linked with precision strike. • Air power is critical to attacking the enemy's deep strike assets and high-value targets. • Tight rules of engagement for centralized strikes and close air support are required. • Air power is used for the suppression of enemy standoff systems to support (complement) ground maneuver. • Air power complicates the enemy's ability to mass and be reinforced. • Air superiority may be contested below 20,000 feet.	• Air power is critical to deterrence achieved through global reach and strike capabilities. • Overhead ISR and signals intelligence are tightly linked with precision strike. • Strategic and operational air mobility and tankers are critical capabilities. • Air and space superiority may be contested at all levels. • Centralized control is critical. • Air power precludes large-scale ground maneuver by the enemy. • Air bases may be contested. • Operations may occur in a nuclear, biological, or chemical environment.

Table 4.4—Continued

	Irregular Warfare	State-Sponsored Hybrid Warfare	Deterrence/Major Combat Operations
The role of ground power	• Ground power is focused on establishing security, obtaining human intelligence, and training indigenous forces. • Maneuver is focused on clearing, holding, and building. • Dispersed operations increase the difficulty of force protection. • Tight rules of engagement demand rigorous target identification.	• Ground power is critical to forcing enemy reaction and to uncovering hidden assets. • Combined-arms operations are fundamental to success. • Ground power closes with enemy forces. • Ground power conducts decentralized operations against dispersed adversaries. • High-intensity MCO-like operations are possible at the brigade level and below. • Lines of communication may be vulnerable.	• Troop deployment is a key signal of national commitment. • Combined-arms operations are the key to success. • Ground maneuver forces an operational reaction from the enemy. • Ground power engages ground units that avoid air attacks and indirect fire. • Ground power is critical for exploiting operational opportunities and pursuing enemy forces. • Ground power deals with hybrid or irregular threats. • Ground power is critical to establishing post-MCO security and stability. • Basing and staging may be contested. • Operations may occur in a nuclear, biological, or chemical environment.
The level of air-ground integration	• Operations are ground-centric but highly dependent on air power. • C2ISR and joint tactical air controllers are best integrated at lower echelons for direct support.	• Balanced operation requires tighter coordination and extensive training and rehearsals. • Integration ensures high responsiveness to ground units and integration at levels below the theater.	• Air power control is highly centralized. • Air superiority is critical to ground maneuver. • Integrated suppression of enemy air defenses is key. • Supported-supporting relationships depend on the operation; the air or ground commander could lead. Leadership could change during an operation.

decision and a lower risk of casualties. What evolved in the IDF was a force very competent in both HIC and LIC.

This all seemed very reasonable at the time.

What Israel encountered in Hezbollah during the Second Lebanon War was a hybrid threat that showed that the IDF had a significant capability gap "in the middle." After the war, the IDF recognized that defeating hybrid adversaries requires joint, combined-arms fire and maneuver skills that it had *consciously* neglected. The IDF focused enormous energy and resources on fixing its deficiencies and, in less than two years, was able to demonstrate in Operation Cast Lead that, at least on a limited scale, it was once again a competent military force across the "rainbow of conflict."

There are similarities between the state of the IDF in 2006 and the U.S. joint force today. After years of protracted irregular warfare in Afghanistan and Iraq, the U.S. military, and particularly the U.S. Army, is stretched to capacity, much like the IDF was during the Second al-Aqsa Intifada. Understandably, the U.S. military is focused on the wars it is currently fighting. U.S. defense officials recognize that preparing for COIN operations has put the force, and particularly the U.S. Army, out of balance. As Stephen Biddle and Jeffrey Friedman write in their assessment of the implications of the Second Lebanon War, deciding to prepare for COIN is a rational decision for the U.S. military:

> Failure in either Iraq or Afghanistan could have grave consequences for U.S. national interests. Until these theaters are stabilized—or unless stability becomes infeasible—it will be essential to maximize U.S. performance in these ongoing wars even if this reduces future potential for some as-yet unseen war elsewhere.[20]

The final point that Biddle and Friedman make is very important in understanding the differences between the U.S. and Israeli strategic situations: Israel does not have the geographic separation from its threats that the United States enjoys. Short-range rockets are not fall-

---

[20] Biddle and Friedman, *The 2006 Lebanon Campaign and the Future of Warfare*, p. xviii.

ing on towns along U.S. borders, but they are a daily threat in Israel. In the United States, this separation from the threat perhaps creates a dangerous illusion that it is reasonable to focus mainly on COIN; this is dangerous because there is a possibility that the irregular opponents the U.S. military is facing in the field could transition to hybrid status with the introduction of an effective standoff weapons capability and a modicum of training. Irregular Taliban adversaries might attain the military capabilities needed to make them hybrid opponents like Hezbollah or Hamas, or at least like the mujahedeen during the Soviet occupation of Afghanistan in the 1980s. Such an evolution could markedly change the operational challenges confronting U.S. forces, particularly because of the United States' reliance on air mobility for operational maneuver and logistics. Thus, Biddle and Friedmans's view that *"it will be essential to maximize U.S. performance in these ongoing wars even if this reduces future potential for some as-yet unseen war elsewhere"* may not be prudent, if the cost of this optimization is neglecting the capabilities needed to prevail against Hezbollah-like adversaries.[21] The IDF understands this. It has not abandoned preparing for the low-end challenges—which it still faces—in order to ensure that it is ready to confront such adversaries as Hezbollah and Hamas. Instead, it now recognizes that, to be a fully useful instrument, it must truly prepare for the full range of military operations.[22]

The ultimate lesson from the Israeli experiences in Lebanon and Gaza is this: An enemy's capabilities will largely determine the war you will have to fight. And, as the IDF learned in Lebanon, it is too late to create an integrated air-ground combined-arms team once you are under fire.

---

[21] Biddle and Friedman, *The 2006 Lebanon Campaign and the Future of Warfare*, p. 89.

[22] Author's discussions with IDF officers, Tel Aviv, February 9–19, 2009. In the IDF, the regional commands focus the units they have assigned on the specific issues they will likely face, e.g., territorial brigades along the West Bank train to deal with irregular adversaries, mostly infiltrators. Nevertheless, this training is built on a base of preparation for high-end, combined-arms operations.

Implications of the IDF Experience

The Israeli experiences in the Second Lebanon War and Gaza have several implications for the U.S. joint force:

- Despite the smaller scale of the conflict, such hybrid opponents as Hezbollah and Hamas constitute a challenge that is qualitatively similar to that posed by major combat operations because of their training, discipline, organization, command and control, and effective standoff weapons (e.g., ATGMs, MANPADS, surface-to-surface rockets). These capabilities are "game changers": Irregular opponents who attain them can rapidly ratchet up the intensity level of a conflict, and defeating these opponents requires different skills than those used in counterinsurgency operations. After years of focusing on LIC operations in Gaza and the West Bank, the IDF (particularly the Israeli Army) was not prepared for the challenges posed by Hezbollah. The U.S. military faces similar issues after years of focusing on irregular warfare in Afghanistan and Iraq.
- There are no single-service solutions to the challenges posed by hybrid adversaries. Israel's training, organizational, and doctrinal changes after the Second Lebanon War, and particularly changes in air-ground-ISR integration, paid off in Operation Cast Lead for the IDF. Similar changes across the doctrine, organization, training, materiel, leadership and education, personnel, and facilities (DOTMLPF) spectrum may be necessary to prepare the U.S. joint force for hybrid opponents.
- Precision standoff fires were critical to—but not sufficient for—coping with hybrid opponents in Lebanon and Gaza, particularly when Israel's adversaries were operating among the population. Thus, as the IDF realized, joint, combined-arms operations are crucial in finding, fixing, and capturing or killing hybrid adversaries who are dispersed and concealed in complex terrain. Furthermore, because of the fleeting nature of the enemy, ground force brigades, rather than higher echelons, were the locus of decentralized tactical decisionmaking and combined-arms action in Gaza.

- Hybrid opponents (e.g., Hamas) become more visible when they take over and move into government buildings; this makes them more susceptible to precision strike. Similarly, intermediate- and long-range rockets are easier to find and destroy because of their size and the requirement that they be launched from relatively open sites. Thus, the IAF was very successful in finding and destroying intermediate- and long-range rockets and in attacking infrastructure targets. This is a unique capability afforded by air power, and it was particularly important in preempting the use of more-capable rockets. Additionally, only fixed-wing aircraft were capable of delivering precision ordnance with the payloads necessary to destroy large, multistory structures and tunnels.

- Persistent ISR coverage is crucial in tracking mobile opponents (particularly mortar, rocket, and ATGM crews) and high-value targets (e.g., intermediate- and long-range rockets in Lebanon, key enemy leaders). UAVs are particularly valuable because of their loitering time and because using unmanned (rather than manned) aircraft in high-threat environments eliminates the risk of losing a pilot. The ability to command and control a combination of ground forces, fixed-wing aircraft, attack helicopters, UAVs, and other assets—all operating off a "common picture" of the battlefield—is critical in attacking fleeting, time-sensitive targets and in avoiding collateral damage.

- Highly dispersed, low-signature targets (e.g., short-range rockets, ATGMs) are difficult to find and attack by air-only means, and attacking them requires comprehensive air-ground-ISR integration at low (i.e., battalion and brigade) levels; common references (e.g., detailed maps and control measures); and a shared, real-time ISR picture.

- Successfully striking targets "amongst the people" in Gaza required a combination of exquisite interagency intelligence, precision strike, and low-yield weapons. These capabilities were needed to identify targets, warn civilians, attack hidden targets (e.g., weapons caches, rockets) and avoid killing the wrong people. This level of intelligence may not be attainable by the Israelis elsewhere (e.g., in Lebanon) or by the United States in current and future conflict environments.

- Armored forces based on tanks and armored personnel carriers are key elements of any force that will fight hybrid enemies with a modicum of training, organization, effective standoff weapons (e.g., ATGMs, MANPADS), IEDs, and mines. Light and medium forces (e.g., Stryker brigade combat teams in the U.S. Army) can complement armored forces, particularly in urban and other complex terrain, but they do not provide the survivability, lethality, or mobility inherent in armored forces. Quite simply, armored forces reduce operational risks and minimize friendly casualties. Information cannot replace armor.

Lessons and Recommendations for the U.S. Army and the U.S. Air Force

In light of these insights, this monograph presents the following lessons and recommendations:

- The skills and processes needed to prevail against hybrid opponents are different in many respects from those required to conduct a counterinsurgency. They require the highly integrated joint, combined-arms fire-and-maneuver skills used in major combat operations, but at a lower organizational level (i.e., the brigade combat team level). The skills and processes needed for integrated air-ground-ISR operations against hybrid adversaries with effective stand-off fires capabilities—skills and processes that may have atrophied among U.S. forces during the United States' protracted counterinsurgency operations in Afghanistan and Iraq—need to be assessed and strengthened in both the U.S. Air Force and the U.S. Army. Therefore, the combat training centers should continue their renewed emphasis on preparing forces for the full range of military operations and should incorporate hybrid operations and hybrid opposing forces into training exercises and experiments.
- U.S. Air Force and U.S. Army training, organization, and equipping efforts need to prepare forces for hybrid challenges, which

could materialize in Afghanistan or elsewhere with the introduction of effective standoff weapons, particularly MANPADS. The appearance of these types of weapons in any numbers in Afghanistan could radically change how U.S. forces operate.

- The inherent nature of hybrid threats requires detailed air-ground-ISR integration, and the U.S. Air Force can make an important contribution by ensuring that it has a highly capable and senior presence in brigade staffs and in subordinate maneuver forces and staffs. In the IDF, C2 during the air-ground phase of Operation Cast Lead was pushed down to the brigade level. Intelligence, fires, and maneuver were fused because of the fleeting nature of targets, the complex operating environment, the dispersed nature of the adversary, and the need to limit collateral damage. It is reasonable to assume that the future hybrid threat environments that the U.S. joint force might face will have similar characteristics. Furthermore, the U.S. Army has adopted a brigade combat team structure not unlike that used by the IDF in Operation Cast Lead. Therefore, the U.S. Air Force and the U.S. Army should assess what kind of air operations planning and C2 capabilities should reside in U.S. Army brigades and consider the integration of capabilities from across the other military services (e.g., the U.S. Navy and U.S. Marine Corps) and federal agencies (e.g., the Central Intelligence Agency, the National Security Agency).

- The threats posed by precision standoff fire systems (e.g., MANPADS, ATGMs) are different than those posed by irregular opponents. Technological countermeasures (e.g., jammers, active armor)—as well as tactics, techniques, and procedures—are needed to defeat these weapons.

- The U.S. joint force needs to develop and institutionalize processes to integrate and control cross-service platforms and capabilities at the level of the brigade combat team. This will be harder in the U.S. joint force than it is in the IDF because, in the latter military, the IAF owns and operates virtually every air platform (i.e., fixed-wing aircraft, rotary-wing aircraft, and most UAVs). However, integration will become more complex in the IDF as UAVs continue to proliferate in the ground forces.

- The U.S. joint force needs to develop common reference systems for urban terrain that enable rapid mutual target recognition, mensuration, and attack. These systems could include predetermined common coordinates and numbering systems for buildings.
- The U.S. joint force needs the capability to find, collaboratively observe, and strike fleeting targets among civilian populations with precision and very-low-yield weapons. This has been done in ad hoc ways in Iraq (in, for example, the 2008 battle in Sadr City) but has not yet been institutionalized (as it was in the IDF before Operation Cast Lead).
- The U.S. Air Force needs to have the capabilities required to destroy large structures (e.g., multistory buildings) and subterranean complexes like those used by Hezbollah and Hamas, but it must also be able to limit civilian casualties and collateral damage. In all likelihood, challenges in this area will increase as adversaries dig deeper and continue to operate in urban areas. Additionally, the realities of fighting in complex terrain, particularly urban areas, can require forces to drop munitions closer to friendly troops. Thus, smaller and highly precise munitions are needed to avoid fratricide.
- To effectively engage hybrid opponents, an air-ground-ISR team must receive detailed training and conduct rehearsals. Therefore, to avoid ad hoc arrangements that limit effectiveness and replicability, the U.S. Air Force and the U.S. Army should examine existing arrangements for joint planning and execution and consider the possibility of establishing habitual relationships between air and ground forces within a theater of operations. As the IDF experience in Gaza shows, trust between air and ground forces is a combat multiplier, and only through habitual association and personal relationships can this trust be truly established.

Final Thoughts

Israel's experiences in Lebanon and Gaza show that hybrid opponents can create significant challenges for nations whose ground forces are focused on irregular warfare and whose air forces are designed to maintain a high-end deterrent and warfighting capability. The Israelis learned the hard way in Lebanon that there was a gap in the IDF's ability to carry out operations "in the middle." As the U.S. joint force prepares to confront the full spectrum of potential future challenges, Israel's experiences are well worth learning from.

Timeline of the 2006 Second Lebanon War

Figure A.1 (on the next page) shows the Winograd Commission's timeline of the Second Lebanon War.

**Figure A.1
Timeline of the Second Lebanon War**

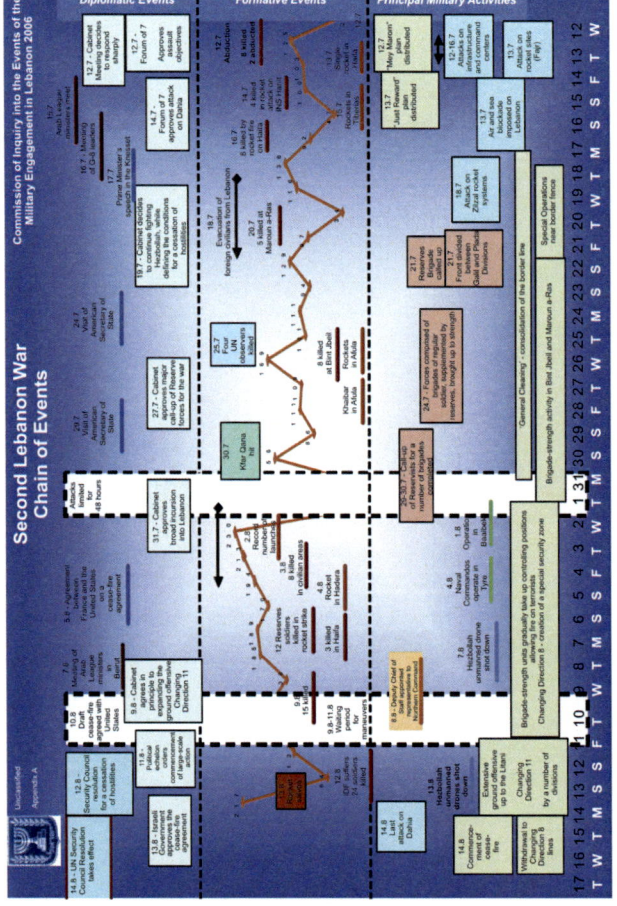

SOURCE: The Winograd Commission, p. 576.
NOTE: The timeline runs from right to left.

RAND MG1085-A.1

Bibliography

Adamsky, Dima, *The Culture of Military Innovation: The Impact of Cultural Factors on the Revolution in Military Affairs in Russia, the US, and Israel*, Stanford, Calif.: Stanford University Press, 2010.

Aguy, Yechezkel, "Mechanized Maneuvers in an Anti Tank and Obstacle Saturated Terrain," presentation at the Maneuver in Complex Terrain Conference, Latrun, Israel, September 1–3, 2009.

Arkin, William M., *Divining Victory: Airpower in the 2006 Israel-Hezbollah War*, Maxwell Air Force Base, Ala.: Air University Press, 2007.

———, "Divine Victory for Whom? Airpower in the 2006 Israel-Hezbollah War," *Strategic Studies Quarterly*, Winter 2007, pp. 98–141.

Bazinet, Kenneth R., and Helen Kennedy, "'An Act of War': Israel Reacts to Slay of 8, Sends Tanks to Lebanon," *New York Daily News*, July 13, 2006.

BBC News, "Rockets Hit Israeli City of Haifa," News.BBC.co.uk, last updated July 13, 2006. As of August 23, 2010:
http://news.bbc.co.uk/2/hi/middle_east/5178058.stm

———, "Who Are Hezbollah?" News.BBC.co.uk, May 21, 2008. As of October 23, 2009:
http://news.bbc.co.uk/2/hi/middle_east/4314423.stm

Ben-David, Alon, "Israeli Armour Fails to Protect MBTs from ATGMs," *Jane's Defence Weekly*, October 4, 2006.

———, "Debriefing Teams Brand IDF Doctrine 'Completely Wrong,'" *Jane's Defence Weekly*, January 3, 2007.

———, "Iranian Influence Looms as Fragile Gaza Ceasefire Holds," *Jane's Defence Weekly*, January 22, 2009.

Berman, Lazar, "Beyond the Basics: Looking Beyond the Conventional Wisdom Surrounding the IDF Campaigns Against Hizbullah and Hamas," SmallWarsJournal.com, April 28, 2011. As of May 31, 2011:
http://smallwarsjournal.com/blog/2011/04/beyond-the-basics/

Biddle, Stephen, and Jeffrey A. Friedman, *The 2006 Lebanon Campaign and the Future of Warfare: Implications for Army and Defense Policy*, Carlisle Barracks, Pa.: U.S. Army War College Strategic Studies Institute, 2008.

Blanford, Nicholas, "Deconstructing Hizbullah's Surprise Military Prowess," *Jane's Intelligence Review*, November 1, 2006.

———, "A Rare Trip Through Hizbullah's Secret Tunnel Network," CSMonitor. com, May 11, 2007. As of August 25, 2009: http://www.csmonitor.com/2007/0511/p01s02-wome.html?page=1

Brom, Shlomo, "Political and Military Objectives in a Limited War Against a Guerilla Organization," in Shlomo Brom and Meir Elran, eds., *The Second Lebanon War: Strategic Perspectives*, Tel Aviv: Institute for National Security, 2007.

Brom, Shlomo, and Meir Elran, eds., *The Second Lebanon War: Strategic Perspectives*, Tel Aviv: Institute for National Security, 2007.

Brun, Itai, "The Second Lebanon War, 2006," unpublished manuscript, undated [2009].

———, "The Second Lebanon War, 2006," in John Andreas Olsen, ed., *A History of Air Warfare*, Washington, D.C.: Potomac Books, 2010.

Byman, Daniel, *Deadly Connections: States that Sponsor Terrorism*, Cambridge, UK: Cambridge University Press, 2005.

———, *A High Price: The Triumphs and Failures of Israeli Counterterrorism*, New York: Oxford University Press, 2011, p. 189.

Byman, Daniel, and Steven Simon, "The No-Win Zone: An After-Action Report from Lebanon," *The National Interest*, No. 86, November/December 2005, pp. 55–61.

Catignani, Sergio, *Israeli Counter-Insurgency and the Intifadas: Dilemmas of a Conventional Army*, London: Routledge, 2008.

———, "Variation on a Theme: Israel's Operation Cast Lead and the Gaza Strip Missile Conundrum," *RUSI Journal*, Vol. 154, No. 4, August 2009, pp. 66–73.

Central Command, Israel Defense Forces, "Washington Institute Briefing," November 1999.

Central Intelligence Agency, "Gaza Strip," *The World Factbook*, no date available. As of February 15, 2011: https://www.cia.gov/library/publications/the-world-factbook/geos/gz.html

Cohen, Stuart, *Israel and Its Army: From Cohesion to Confusion*, London: Routledge, 2008.

Cohen, Yoram, and Jeffrey White, *Hamas in Combat: The Military Performance of the Palestinian Islamic Resistance Movement*, Policy Focus No. 97, Washington, D.C.: The Washington Institute for Near East Policy, 2009.

The Commission for the Examination of the Events of the 2006 Campaign in Lebanon, *The Second Lebanon War: Final Report*, Vol. I, January 2008.

Cordesman, Anthony H., "The 'Gaza War': A Strategic Analysis," draft report, February 2, 2009.

Cordesman, Anthony, George Sullivan, and William Sullivan, *Lessons of the 2006 Israeli-Hezbollah War*, Washington, D.C.: The Center for Strategic and International Studies, 2007.

Council on Foreign Relations, "Hamas," web page, updated August 27, 2009. As of November 17, 2009:
http://www.cfr.org/publication/8968

———, "Hezbollah," web page, updated July 15, 2010. As of March 1, 2011:
http://www.cfr.org/publication/9155/hezbollah_aka_hizbollah_hizbullah.html

Crooke, Alastair, and Mark Perry, "How Hezbollah Defeated Israel: Part 1—Winning the Intelligence War," ATimes.com, October 12, 2006. As of March 29, 2010:
http://www.atimes.com/atimes/Middle_East/HJ12Ak01.html

———, "How Hezbollah Defeated Israel: Part 2—Winning the Ground War," ATimes.com, October 13, 2006. As of March 29, 2010:
http://www.atimes.com/atimes/Middle_East/HJ13Ak01.html

Dalton, L. Craig, *Systemic Operational Design: Epistemological Bumpf or the Way Ahead for Operational Design?* Fort Leavenworth, Kan.: School of Advanced Military Studies, U.S. Army Command and General Staff College, 2006.

Defense Industry Daily, "2007–08: Israel Looks to Replenish Bomb Stocks," DefenseIndustryDaily.com, August 7, 2007. As of October 6, 2009:
http://www.defenseindustrydaily.com/
israel-looks-to-replenish-bomb-stocks-03590/

Devenny, Patrick, "Hezbollah's Strategic Threat to Israel," *Middle East Quarterly*, Vol. XIII, No. 1, Winter 2006, pp. 31–38.

Donley, Michael B., "Air Force Modernization and Recapitalization Strategy," keynote address prepared for the Aerospace Industries Association, November 18, 2008.

Donley, Michael B., and Norton A. Schwartz, *United States Air Posture Statement: 2009*, May 19, 2011.

Dudney, Robert S., "The Air War over Hezbollah," *Air Force Magazine*, Vol. 89, No. 9, September 2006.

Dunnigan, James, "Hapless Hezbollah ATGMs Revealed," StrategyPage.com, September 7, 2008. As of September 20, 2010:
http://www.strategypage.com/dls/articles/20089721428.asp

Eiland, Giora, "The Decision Making Process in Israel," in Shlomo Brom and Meir Elran, eds., *The Second Lebanon War: Strategic Perspectives*, Tel Aviv: Institute for National Security Studies, 2007.

———, "The IDF in the Second Intifada," *Strategic Assessment*, Vol. 13, No. 3, October 2010, pp. 27–38.

Eldar, Akiva, "Jerusalem Demolitions May Spark Repeat of 1996 Riots," Haaretz.com, last updated October 3, 2009. As of April 21, 2011: http://www.haaretz.com/print-edition/features/ jerusalem-demolitions-may-spark-repeat-of-1996-riots-1.271808

Elkus, Adam, "The Hezbollah Myth and Asymmetric Warfare," SmallWarsJournal.com, August 17, 2010. As of August 5, 2011: http://smallwarsjournal.com/blog/journal/docs-temp/497-elkus.pdf

Erlanger, Steven, "A Gaza War Full of Traps and Trickery," NYTimes.com, January 11, 2009. As of February 2, 2011: http://www.nytimes.com/2009/01/11/world/middleeast/11hamas.html?_r=1&part ner=permalink&exprod=permalink

Eshel, Amir, "The IAF Doctrine of Counter-Terror Air Warfare," in The Fisher Brothers Institute for Air and Space Strategic Studies, *Air Power Against Terrorism, The Fisher Brothers Institute International Conference, May 2005*, Herzliya, Israel, 2005.

Eshel, David, "Lebanon 2006," *Tank Magazine*, December 2006.

———, "Hezbollah Anti-Amour Tactics and Weapons: Assessment of the Second Lebanon War," Defense-Update.com, 2007. As of September 20, 2010: http://defense-update.com/analysis/lebanon_war_4.htm

———, "Hezbollah's Intelligence War: Assessment of the Second Lebanon War," Defense-Update.com, 2007. As of September 20, 2010: http://defense-update.com/analysis/lebanon_war_1.htm

———, "New Tactics Yield Solid Victory in Gaza," AviationWeek.com, March 11, 2009. As of October 4, 2009: http://www.aviationweek.com/aw/generic/story_channel. jsp?channel=defense&id=news/GAZA031109.xml#

Evron, Yair, "Deterrence and Its Limitation," in Shlomo Brom and Meir Elran, eds., *The Second Lebanon War: Strategic Perspectives*, Tel Aviv: Institute for National Security Studies, 2007.

"Excerpts from the Winograd Report," YNetNews.com, April 30, 2007. As of August 30, 2011: http://www.ynetnews.com/articles/0,7340,L-3394059,00.html

Exum, Andrew, *Hizballah at War: A Military Assessment*, Policy Focus No. 63, Washington, D.C.: The Washington Institute for Near East Policy, 2006.

Farquhar, Scott C., ed., *Back to Basics: A Study of the Second Lebanon War and Operation CAST LEAD*, Fort Leavenworth, Kan.: Combat Studies Institute Press, 2009.

Federation of American Scientists, "Man-Portable Air Defense System (MANPADS) Proliferation: Understanding the Problem," web page, undated. As of July 29, 2010:
http://www.fas.org/programs/ssp/asmp/MANPADS.html

Feldman, Yotam, "Dr. Naveh, or, How I Learned to Stop Worrying and Walk Through Walls," Haaretz.com, last updated October 25, 2007. As of September 21, 2009:
http://www.haaretz.com/hasen/spages/917158.html

Friedman, Ina, "Moral Morass," *The Jerusalem Report*, September 4, 2006, p. 12.

Gentile, Gian P., "The Imperative for an American General Purpose Army That Can Fight," *Orbis*, Vol. 53, No. 3, Summer 2009.

Geren, Pete, and George W. Casey, *2009 Army Posture Statement*, May 2009.

Gil, Avi, "Operation 'Cast Lead'—Gaza: Insight and Lessons Learned from 'Al-Atatra' Battle—the Paratroopers Brigade," briefing, RAND Corporation, Arlington, Va., April 1, 2009.

Glenn, Russell W., *Glory Restored: The Implications of the 2008–2009 Gaza War in Times of Extended Conflict*, August 25, 2010.

GlobalSecurity.org, "HAMAS Rockets," web page, date not available. As of September 25, 2009:
http://www.globalsecurity.org/military/world/para/hamas-qassam.htm

———, "Hizballah Rockets," web page, date not available. As of February 1, 2011:
http://www.globalsecurity.org/military/world/para/hizballah-rockets.htm

Goldstone, Richard, "Reconsidering the Goldstone Report on Israel and War Crimes," WashingtonPost.com, April 1, 2011. As of April 2, 2011:
http://www.washingtonpost.com/opinions/reconsidering-the-goldstone-report-on-israel-and-war-crimes/2011/04/01/AFg111JC_story.html

Gordon, James, "Iran Called Source of Missile that Struck Ship," *New York Daily News*, July 19, 2006, p. 19.

Grau, Lester W., and Michael A. Gress, eds., trans., *The Soviet Afghan-War: How a Superpower Fought and Lost*, Lawrence, Kan.: University of Kansas Press, 2002.

Greenberg, Hanan, "24 Troops Killed in Day of Battles," YNetNews.com, last updated August 13, 2006. As of March 30, 2010:
http://www.ynetnews.com/articles/0,7340,L-3290185,00.html

———, "Brigade Commander: Hamas Will Draw Lessons, Grow Stronger," YNetNews.com, last updated January 23, 2009. As of January 17, 2011:
http://www.ynetnews.com/articles/0,7340,L-3660868,00.html

Gupta, Dipak K., and Kusum Mundra, "Suicide Bombing as a Strategic Weapon," *Terrorism and Political Violence*, Vol. 17, 2005.

Harel, Amos, and Avi Issacharoff, *34 Days: Israel, Hezbollah, and the War in Lebanon*, New York: Palgrave Macmillan, 2008.

Harel, Amos, Avi Issacharoff, and Haaretz Correspondent, "Analysis: A Hard Look at Hamas' Capabilities," Haaretz.com, December 26, 2008. As of September 1, 2011:
http://www.haaretz.com/print-edition/news/
analysis-a-hard-look-at-hamas-capabilities-1.260281

Harel, Dan, "The Fire Delivery Concept in Operation 'Cast Lead,'" briefing, Israel Fire and Combined Arms in Urban Terrain Conference, November 8–11, 2010.

———, "81% of the Fire—Precision Guided Munitions," *Israel Defense*, Vol. 1, February–March 2011, pp. 18–21.

Hart, B. H. Liddell, *The Rommel Papers*, New York: Harcourt, Grace, 1953.

Hendel, Yoaz, "Failed Tactical Intelligence in the Lebanon War," *Strategic Assessment*, Vol. 9, No. 3, November 2006.

"Hezbollah Leader Calls for Muslim-Christian Coexistence," Lebanese Broadcasting Corporation, May 26, 2000, in *BBC Summary of World Broadcasts*, May 27, 2000.

Hirsch, Moshe, "Legislation Note: Treaty-Making Power: Approval of the Israel-Egypt 'Philadelphi Accord' by the Knesset," *Israel Law Review*, Winter 2006.

Hoffman, Frank G., *Conflict in the 21st Century: The Rise of Hybrid Wars*, Arlington, Va.: Potomac Institute for Policy Studies, 2007.

———, "Hybrid vs. Compound War: The Janus Choice—Defining Today's Multifaceted Conflict," *Armed Forces Journal*, October 2009.

Hosmer, Stephen T., *The Conflict Over Kosovo: Why Milosevic Decided to Settle When He Did*, Santa Monica, Calif.: RAND Corporation, MR-1351-AF, 2001. As of August 5, 2011:
http://www.rand.org/pubs/monograph_reports/MR1351.html

Hoteit, Amine, "Harb Israeil ala Gaze wa Asaraha ala al-Istrategiya al-Aaskariya [The Gaza War and Its Impact on the Military Strategy]," Al-Jazeera Center for Studies, undated.

Hroub, Khaled, *HAMAS: Political Thought and Practice*, Washington, D.C.: Institute for Palestine Studies, 2000.

Huber, Thomas M., "Huber Comments on Hybrid Warfare and Compound Warfare for DMH Faculty and Others Interested," post on the Combined Arms Center Blog, February 12, 2009. As of September 27, 2010: http://usacac.army.mil/BLOG/blogs/hist/archive/2009/02/12/huber-comments-on-hybrid-warfare-and-compound-warfare-for-dmh-faculty-and-others-interested. aspx

Inbar, Efraim, "How Israel Bungled the Second Lebanon War," *Middle East Quarterly*, Summer 2007.

Intelligence and Terrorism Information Center at the Israel Intelligence Heritage & Commemoration Center, *Hamas's Military Buildup in the Gaza Strip*, April 8, 2008.

International Institute for Strategic Studies, *The Military Balance 2005–2006*, London, 2005.

———, *The Military Balance 2007*, London, 2007.

———, *The Military Balance 2009*, Milton Park, UK: Routledge, 2009.

Isherwood, Michael W., *Airpower for Hybrid Warfare*, Mitchell Paper No. 3, Arlington, Va.: Mitchell Institute Press, 2009.

"Islamic Jihad Armed Wing Rejects Abbas Calls to Halt Rocket Fire," *BBC Monitoring Middle East*, June 23, 2006.

IslamicNews.net, "Khottat al-Moqawama fi harb Gaza . . . Istrategiya Mahaliya [The Resistance Plan During the Gaza War . . . a Local (i.e., National) Strategy]," undated. As of August 6, 2009: http://www.islamicnews.net/Document/ShowDoc08.asp?DocID=127401&TypeID=8&TabIndex=2

Israel Defense Forces, "IDF Press Releases," web page, undated. As of August 9, 2011: http://dover.idf.il/IDF/English/opcast/op/press/default.htm

———, "Operation Cast Lead Against Terror Infrastructure in Gaza Continues," press release, December 27, 2008. As of September 24, 2009: http://dover.idf.il/IDF/English/opcast/op/press/2702.htm

———, "Rocket Statistics, 3 Jan 2009," web page, January 3, 2009. As of January 31, 2011: http://idfspokesperson.com/2009/01/03/rocket-statistics-3-jan-2009/

———, "Precision Airstrikes on Hamas Terror Targets 7 Jan. 2009," web page, January 7, 2009. As of August 9, 2011: http://idfspokesperson.com/2009/01/07/precision-airstrikes-on-hamas-terror-targets-7-jan-2009/

Israel Ministry of Foreign Affairs, "Victims of Palestinian Violence and Terrorism Since September 2000," web page, undated. As of September 19, 2009:
http://www.mfa.gov.il/MFA/Terrorism-+Obstacle+to+Peace/
Palestinian+terror+since+2000/Victims+of+Palestinian+Violence+and+Terrorism+s
inc.htm

———, "The IDF's Secret Weapon Against Hizbullah," article, December 5, 1996. As of November 17, 2009:
http://www.mfa.gov.il/MFA/Archive/Articles/1996/THE%20IDF-S%20
SECRET%20WEAPON%20AGAINST%20HIZBULLAH%20-%2005-Dec

———, "Operation Defensive Shield: Special Update," web page, March 29, 2002. As of February 1, 2011:
http://www.mfa.gov.il/MFA/MFAArchive/2000_2009/2002/3/Operation%20
Defensive%20Shield

———, "Winograd Commission Submits Interim Report," web page, April 30, 2007. As of March 1, 2009:
http://www.mfa.gov.il/MFA/Government/Communiques/2007/Winograd+Inquir
y+Commission+submits+Interim+Report+30-Apr-2007.htm

———, "The Operation in Gaza—Factual and Legal Aspects," web page, July 29, 2009. As of February 2, 2011:
http://www.mfa.gov.il/MFA/Terrorism-+Obstacle+to+Peace/
Hamas+war+against+Israel/Operation_in_Gaza-Factual_and_Legal_Aspects.htm

———, *Initial Response to Report of the Fact Finding Mission on Gaza Established Pursuant to Resolution S-9/1 of the Human Rights Council*, September 24, 2009.

"Israel Warns Hizbullah War Would Invite Destruction," Reuters, March 10, 2008.

Isseroff, Ami, "Biography of Ehud Olmert," Zionism-Israel.com, updated September 2008. As of November 23, 2009:
http://www.zionism-israel.com/bio/Olmert_biography.htm

Jaber, Hala, *Hezbollah: Born With a Vengeance*, New York: Columbia University Press, 1997.

Jacobs, Jody, David E. Johnson, Katherine Comanor, Lewis Jamison, Leland Joe, and David Vaughan, *Enhancing Fires and Maneuver Capability Through Greater Air-Ground Joint Interdependence*, Santa Monica, Calif.: RAND Corporation, MG-793-AF, 2009. As of August 5, 2011:
http://www.rand.org/pubs/monographs/MG793.html

Jewish Daily Report, "Israel Prepares for War with Lebanon/Hizbollah Terrorists," JewishDailyReport.wordpress.com, July 23, 2010. As of September 20, 2010:
http://jewishdailyreport.wordpress.com/2010/07/23/
israel-prepares-for-war-with-lebanonhizbollah-terrorists

Jewish Virtual Library, "Statistics on Operation 'Defensive Shield,'" web page, undated. As of February 15, 2011:
http://www.jewishvirtuallibrary.org/jsource/History/defensiveshield.html

Johnson, David E., *Learning Large Lessons: The Evolving Roles of Ground Power and Air Power in the Post-Cold War Era*, Santa Monica, Calif.: RAND Corporation, MG-405-1-AF, 2007. As of August 5, 2011:
http://www.rand.org/pubs/monographs/MG405-1.html

———, *Military Capabilities for Hybrid War: Insights from the Israel Defense Forces in Lebanon and Gaza*, Santa Monica, Calif.: RAND Corporation, OP-285-A, 2010. As of August 9, 2011:
http://www.rand.org/pubs/occasional_papers/OP285.html

———, "Iran's Counter-Strike," *Providence Journal*, February 25, 2010.

Johnson, David E., and John Gordon IV, *Observations on Recent Trends in Armored Forces*, Santa Monica, Calif.: RAND Corporation, OP-287-A, 2010. As of August 9, 2011:
http://www.rand.org/pubs/occasional_papers/OP287.html

Johnson, David E., Adam Grissom, and Olga Oliker, *In the Middle of the Fight: An Assessment of Medium-Armored Forces in Past Military Operations*, Santa Monica, Calif.: RAND Corporation, MG-709-A, 2008. As of August 8, 2011:
http://www.rand.org/pubs/monographs/MG709.html

Johnson, David E., Jennifer D. P. Moroney, Roger Cliff, M. Wade Markel, Laurence Smallman, and Michael Spirtas, *Preparing and Training for the Full Spectrum of Military Challenges: Insights from the Experiences of China, France, the United Kingdom, India, and Israel*, Santa Monica, Calif.: RAND Corporation, MG-836-OSD, 2009. As of September 21, 2011:
http://www.rand.org/pubs/monographs/MG836.html

Johnson, David E., Karl P. Mueller, and William H. Taft, *Conventional Coercion Across the Spectrum of Operations: The Utility of U.S. Military Forces in the Emerging Security Environment*, Santa Monica, Calif.: RAND Corporation, MR-1494-A, 2003. As of August 8, 2011:
http://www.rand.org/pubs/monograph_reports/MR1494.html

Kainikara, Sanu, and Russell Parking, *Pathways to Victory: Observations from the 2006 Israel-Hezbollah Conflict*, Tuggeranong, Australia: Air Power Development Center, 2007.

Katz, Yaakov, "Defense Officials Debate Peretz's Rumored Appointment," *The Jerusalem Post*, April 5, 2006, p. 2.

———, "IDF Set for Massive Assault on Lebanon," *Jerusalem Post*, July 13, 2006, p. 1.

———, "Wadi Saluki Battle—Microcosm of the War's Mistakes," *The Jerusalem Post*, August 29, 2006.

———"IDF Readying for Gaza Incursion—but Not Yet," *The Jerusalem Post*, September 6, 2007, p. 3.

———, "Security and Defense: The Story of 'Changing Direction 11,'" JPost.com, October 1, 2008. As of September 7, 2010: http://fr.jpost.com/servlet/Satellite?apage=1&cid=1199964893710&pagename=JPost/JPArticle/ShowFull

———, "Future Battlefield Will Be More Lethal," JPost.com, September 2, 2009. As of September 4, 2009: http://fr.jpost.com/servlet/Satellite?cid=1251804474590&pagename=JPArticle%2FShowFull

———, "Small Terror Group Said Behind Katyushas Fired at North," JPost.com, September 11, 2009. As of September 30, 2009: http://www.jpost.com/Israel/Article.aspx?id=154587

———, "IDF Declassifies Intelligence on Hizbullah's Southern Lebanon Deployment," *Jane's Defence Weekly*, July 9, 2010.

Katz, Yaakov, and Sam Ser, "IDF Report Card," *The Jerusalem Post*, August 25, 2006, p. 16.

Kershner, Isabel, "Israeli Military Says Its Actions in Gaza War Did Not Violate International Law," *The New York Times*, April 23, 2009, p. A10.

Kessner, B. C., "New Department Transforming IAF for LIC Success," *Defense Daily International*, Vol. 6, No. 10, March 11, 2005.

Keymer, Eleanor, *Jane's World Armies, Issue Twenty-Seven*, Coulsdon, UK: HIS, 2010.

Kober, Avi, "The Israel Defense Forces in the Second Lebanon War: Why the Poor Performance?" *Journal of Strategic Studies*, Vol. 31, No. 1, February 2008.

———, *Israel's Wars of Attrition: Attrition Challenges to Democratic States*, London: Routledge, 2009.

Kraft, Dina, and David Blair, "Israeli Army Admits 189 Children Were Killed in Gaza," *The Daily Telegraph*, March 26, 2009, p. 19.

Lambeth, Benjamin S., *Air Operations in Israel's War Against Hezbollah: Learning from Lebanon and Getting It Right in Gaza*, Santa Monica, Calif.: RAND Corporation, MG-835-AF, 2011. As of August 8, 2011: http://www.rand.org/pubs/monographs/MG835.html

Lappin, Yaakov, "IDF Releases Cast Lead Casualty Numbers," JPost.com, March 26, 2009. As of September 25, 2009: http://www.jpost.com/Israel/Article.aspx?id=137286

Lee, Jesse, "A New Strategy for Afghanistan and Pakistan," post on the White House Blog, March 27, 2009. As of August 24, 2009:
http://www.whitehouse.gov/
blog/09/03/27/A-New-Strategy-for-Afghanistan-and-Pakistan/

Levinson, Charles, "Israel's Ground Assault Marks Shift in Strategy," Online.WSJ.com, January 5, 2009. As of October 5, 2009:
http://online.wsj.com/article/SB123106067991451749.html

Lewy, Guenter, *America in Vietnam*, Oxford: Oxford University Press, 1978.

Luttwak, Edward, "In Praise of Aerial Bombing," ForeignPolicy.com, March/April 2010. As of June 1, 2010:
http://www.foreignpolicy.com/articles/2010/02/22/in_praise_of_aerial_bombing

Lynch, Colum, "U.N. Voices 'Extreme Shock' over Israeli Strike," *The Washington Post*, July 30, 2006, p. A9.

MacFarland, Sean, Michael Shields, and Jeffrey Snow, "White Paper for CSA: The King and I—The Impending Crisis in Field Artillery's Ability to Provide Fire Support to Maneuver Commanders," undated [2008].

al-Madhoune, Ibrahim, "Israeil Hinama la Tantaser Touhzam wa al-Mouqawama hinama la Touhzam, Tantaser [Israel: Where It Does Not Win, It Is Defeated; The Resistance: Where It Is Not Defeated, It Wins]," The Palestinian Information Center, March 3, 2009. As of July 10, 2009:
http://www.palestine-info.info/ar/default.aspx?xyz=U6Qq7k%2bcOd87MDI46m
9rUxJEpMO%2bi1s7caQUZhtjLgUSlAWwNWH4mURQnRZpLo8FBlPxNy1%
2b%2fzq9L8aHiKHtdWjcwspHpiAJ0zBWpUBfVo%2bIWGzwl0GiBZTpkQ0
5Rz35xc1lCc9vh2w%3d

Mahmoud, Arwa, "Kital Hezbollah, al-Din fi mouwajahat Isra'il, kifa intasar Hezbollah fi harb tamouz 2006 [Hezbollah's Fight, Religion in the Confrontation with Israel, How Hezbollah Won in the July 2006 War]," Amir Publication, 2008.

Mahnaimi, Uzi, "Humbling of the Supertroops Shatters Israeli Army Morale," TimesOnline.co.uk, August 27, 2006. As of August 31, 2010:
http://www.timesonline.co.uk/tol/news/world/article620874.ece

Makovsky, David, and Jeffrey White, *Lessons and Implications of the Israel-Hizballah War: A Preliminary Assessment*, Policy Focus No. 60, Washington, D.C.: The Washington Institute for Near East Policy, 2006.

Manning, Scott E., "'The Second War in Lebanon,' Israel-Hezbollah War of 2006: Was Air Power Able to Accomplish the Goals of the Israeli Government?" thesis, Maxwell Air Force Base, Ala.: Air University, 2008.

Matthews, Matt M., "Interview with BG (Ret.) Shimon Naveh, 1 November 2007," U.S. Army Combined Arms Center Combat Studies Institute, Fort Leavenworth, Kan., undated [2007].

———, *We Were Caught Unprepared: The 2006 Hezbollah-Israeli War*, The Long War Series Occasional Paper No. 26, Fort Leavenworth, Kan.: U.S. Army Combined Arms Center Combat Studies Institute, 2006.

———, "Hard Lessons Learned: A Comparison of the 2006 Hezbollah-Israeli War and Operation CAST LEAD: A Historical Overview," in Scott C. Farquhar, ed., *Back to Basics: A Study of the Second Lebanon War and Operation CAST LEAD*, Fort Leavenworth, Kan.: Combat Studies Institute Press, 2009.

McChrystal, Stanley A., "COMISAF's Initial Assessment," Headquarters, International Security Assistance Force, Kabul, Afghanistan, August 30, 2009.

McDermott, Roger N., "Russia's Conventional Armed Forces and the Georgian War," *Parameters*, Spring 2009, pp. 65–80.

The Meir Amit Intelligence and Terrorism Information Center, *Fathi Hamad, the Hamas Administration's Interior Minister, Revealed that as Many as 700 Hamas Military-Security Operatives Were Killed During Operation Cast Lead*, November 3, 2010.

Merom, Gil, "The Second Lebanon War: Democratic Lessons Imperfectly Applied," *Democracy and Security*, Vol. 4, No. 1, January 2008, pp. 5–33.

Myre, Greg, "As Israelis Pull Out, the Question Lingers: Who'll Control Gaza?" *The New York Times*, September 11, 2005, p. 8.

Myre, Greg, and Steven Erlanger, "Clashes Spread to Lebanon as Hezbollah Raids Israel," *The New York Times*, July 12, 2006.

"Nasrallah Admits 'Intelligence Error,'" *The Jerusalem Post*, February 3, 2007.

Noe, Nicholas, "A Response to Andrew Exum's 'Hizbollah at War: A Military Assessment,'" Mideastwire.com, undated. As of May 25, 2010: http://www.mideastwire.com/downloads/Response%20to%20Andrew%20Exum.pdf

Norton, Augustus Richard, *Hezbollah*, Princeton: Princeton University Press, 2007.

Olsen, John Andreas, ed., *A History of Air Warfare*, Washington, D.C.: Potomac Books, 2010.

Opall-Rome, Barbara, "In Gaza, Both Sides Reveal New Gear," DefenseNews.com, January 5, 2009. As of September 24, 2009: http://www.defensenews.com/story.php?i=3885990

———, "Maj. Gen. Ido Nehushtan: Commander, Israel Air & Space Force," DefenseNews.com, August 3, 2009. As of August 10, 2009: http://www.defensenews.com/story.php?i=4216231

———, "Israel Blurs Roles, Missions in Ground War Concept," DefenseNews.com, October 25, 2010. As of August 8, 2011: http://www.defensenews.com/story.php?i=4965041&c=FEA&s=SPE

"Operation Defensive Shield (2002)," YNetNews.com, March 12, 2009. As of September 22, 2009: http://www.ynetnews.com/articles/0,7340,L-3685678,00.html

Oren, Amir, "IDF Girds for Possibility of War with Syria, Hezbollah in 2007," Haaretz.com, last updated June 11, 2006. As of March 1, 2009: http://www.haaretz.com/hasen/spages/784053.html

Owen, Robert C., and Karl P. Mueller, *Airlift Capabilities for Future U.S. Counterinsurgency Operations*, Santa Monica, Calif.: RAND Corporation, MG-565-AF, 2007. As of August 8, 2011: http://www.rand.org/pubs/monographs/MG565.html

Owen, William F., "The War of New Words: Why Military History Trumps Buzzwords," *Armed Forces Journal*, November 2009.

Palestinian Centre for Human Rights, "Confirmed Figures Reveal the True Extent of the Destruction Inflicted upon the Gaza Strip," press release, March 12, 2009. As of September 25, 2009: http://www.pchrgaza.org/portal/en/index.php?option=com_content&view=a rticle&id=1073:confirmed-figures-reveal-the-true-extent-of-the-destruction- inflicted-upon-the-gaza-strip-israels-offensive-resulted-in-1417-dead-including- 926-civilians-255-police-officers-and-236-fighters&catid=36:pchrpressreleases&It emid=194

"Palestinian Groups Claim Rocket Attacks Against Israeli Targets," *BBC Monitoring Middle East*, June 22, 2006.

"Palestinian Weapons Deployed Against Israel During Operation Cast Lead," *Journal of Palestine Studies*, Vol. 38, No. 3, Spring 2009, pp. 192–200.

Parton, Neville, "Israel's 2006 Campaign in Lebanon: A Failure of Air Power or a Failure of Doctrine?" *Royal Air Force Air Power Review*, Vol. 10, No. 2, Summer 2007, pp. 80–91.

Peri, Yoram, *Generals in the Cabinet Room: How the Military Shapes Israeli Policy*, Washington, D.C.: United States Institute of Peace Press, 2006.

Pfeffer, Anshel, "After Maroun al-Ras Battle, Bint-Jbail Looms as Next Challenge," *The Jerusalem Post*, July 24, 2009.

al-Qassam, "Penetrating Arrow Security Operation," web page, undated. As of August 6, 2009: http://www.alqassam.ps/arabic/operations2.php?id=28

Qassem, Naim, *Hizbullah: The Story from Within*, Dalia Khalil, trans., London: SAQI, 2005.

Ravid, Barak, "Olmert: Gaza War Won't End Until Rockets and Smuggling Stop," Haaretz.com, January 12, 2009. As of September 1, 2009: http://www.haaretz.com/hasen/spages/1054800.html

Romm, Giora, "A Test of Rival Strategies: Two Ships Passing in the Night," in Shlomo Brom and Meir Elran, eds., *The Second Lebanon War: Strategic Perspectives*, Tel Aviv: Institute for National Security Studies, 2007.

Rose, David, "The Gaza Bombshell," *Vanity Fair*, April 2008.

Rubin, Uzi, *The Rocket Campaign Against Israel During the 2006 Lebanon War*, Ramat Gan, Israel: The Begin-Sadat Center for Strategic Studies, Bar-Ilan University, 2007.

Schanzer, Jonathan, *HAMAS vs. FATAH: The Struggle for Palestine*, New York: Palgrave Macmillan, 2008.

Schneider, Howard, "Israel's Missile Defense System Is Progressing Steadily," WashingtonPost.com, September 19, 2009. As of September 30, 2009: http://www.washingtonpost.com/wp-dyn/content/article/2009/09/18/AR2009091801787.html?sid=ST2009091701841

Shamir, Eitan, *Transforming Command: The Pursuit of Mission Command in the U.S., British, and Israeli Armies*, Stanford: Stanford University Press, 2011.

Shamir, Eitan, and Uzi Ben-Shalom, "Mission Command Philosophy from Theory to Practice: The IDF Case," in Military Psychology Center, Ground Forces (IDF), "Abstracts: The 3rd International Military Psychology Conference in Israel," undated [February 2008].

Shapir, Yiftah, "Hamas' Weapons," *Strategic Assessment*, Vol. 11, No. 4, February 2009.

Sharp, Jeremy M., Christopher Blanchard, Kenneth Katzman, Carol Migdalovitz, Alfred Prados, Paul Gallis, Dianne Rennack, John Rollins, Steve Bowman, and Connie Veillette, *Lebanon: The Israel-Hamas-Hezbollah Conflict*, Washington, D.C.: Congressional Research Service, Library of Congress, 2006.

Siboni, Gabriel, "The Military Campaign in Lebanon," in Shlomo Brom and Meir Elran, eds., *The Second Lebanon War: Strategic Perspectives*, Tel Aviv: Institute for National Security Studies, 2007.

Smith, Rupert, *The Utility of Force: The Art of War in the Modern World*, New York: Vintage Books, 2005.

Spencer, James, "Intelligence Lessons from Hizbullah's Ground Campaign in 2006," *The British Army Review*, No. 148, Winter 2009/2010, pp. 96–105.

The State of Israel, "Ehud Olmert," web page, 2009. As of November 23, 2009: http://www.knesset.gov.il/lexicon/eng/olmert_ehud_eng.htm

Storr, Jim, "Reflections on the War in Lebanon," *The RUSI Journal*, Vol. 152, No. 2, April 2007, pp. 70–72.

Susser, Leslie, "Marching to the Livni Beat," *The Jerusalem Post*, July 7, 2008, p. 14.

Tira, Ron, *The Limitations of Standoff Firepower–Based Operations: On Standoff Warfare, Maneuvers, and Decision*, Tel Aviv: Institute for National Security Studies, 2007.

———, *The Nature of War: Conflicting Paradigms and Israeli Military Effectiveness*, Eastbourne, UK: Sussex Academic Press, 2010.

———, "Israel's Strategy (or Lack of) Towards Iran's Forward Rocket Deployments in Lebanon and Gaza," *Infinity Journal*, No. 1, Winter 2010, pp. 14–17.

U.S. Department of Defense, *Irregular Warfare*, DoDD 3000.07, December 1, 2008.

U.S. Department of State, "Foreign Terrorist Organizations," web page, November 24, 2010. As of February 31, 2011:
http://www.state.gov/s/ct/rls/other/des/123085.htm

U.S. Department of the Air Force, *Air Force Basic Doctrine*, AFDD 1, Washington, D.C., 2003.

———, *Irregular Warfare*, AFDD 2-3, Washington, D.C., 2007.

U.S. Department of the Army, *2008 Army Posture Statement: A Campaign Quality Army with Joint and Expeditionary Capabilities*, Washington, D.C., 2008.

U.S. Department of the Army and Marine Corps Combat Development Command, U.S. Department of the Navy, *Counterinsurgency*, FM 3-24/MCWP 3-33.5, 2006.

United Nations, *Illegal Israeli Actions in Occupied East Jerusalem and the Rest of the Occupied Palestinian Territory: Report of the Secretary-General Prepared Pursuant to General Assembly Resolution ES-10/10*, July 30, 2002.

United Nations Human Rights Council, *Human Rights in Palestine and Other Occupied Arab Territories: Report of the United Nations Fact Finding Mission on the Gaza Conflict*, A/HRC/12/48, September 15, 2009.

United Nations Office for the Coordination of Humanitarian Affairs, "Gaza Situation Map," May 2006.

United Nations Security Council, "Security Council Stresses Urgent Need for Humanitarian Access to Palestinians, Welcomes Fact-Finding Team to Examine Events at Jenin Refugee Camp," press release SC/7369, April 19, 2002. As of February 15, 2011:
http://www.un.org/News/Press/docs/2002/SC7369.doc.htm

———, "Security Council Calls for End to Hostilities Between Hizbollah, Israel, Unanimously Adopting Resolution 1701," press release SC/8088, August 11, 2006. As of August 31, 2010:
http://www.un.org/News/Press/docs/2006/sc8808.doc.htm

UNWatch.org, "U.K. Commander Tells UN Council: 'IDF Took More Precautions Than Any Military in History of Warfare,'" web page, October 16, 2009. As of February 2, 2011:
http://www.unwatch.org/site/c.bdKKISNqEmG/b.1289203/apps/s/content.asp?ct=7591585

van Creveld, Martin, "Israel's Lebanese War: A Preliminary Assessment," *The RUSI Journal*, Vol. 151, No. 5, October 2006, pp. 40–43.

———, "Israel's War with Hezbollah Was Not a Failure," Forward.com, January 30, 2008. As of May 28, 2010:
http://www.forward.com/articles/12579/

Vego, Milan N., "A Case Against Systemic Operational Design," *Joint Force Quarterly*, No. 53, Second Quarter 2009, pp. 69–75.

Vick, Alan J., Adam Grissom, William Rosenau, Beth Grill, and Karl P. Mueller, *Air Power in the New Counterinsurgency Era: The Strategic Importance of USAF Advisory and Assistance Missions*, Santa Monica, Calif.: RAND Corporation, MG-509-AF, 2006. As of August 8, 2011:
http://www.rand.org/pubs/monographs/MG509.html

von Clausewitz, Carl, *On War*, Michael Howard and Peter Paret, eds., trans., Princeton: Princeton University Press, 1984.

Waits, Gidi, and Uri Blau, "General Mofaz Runs for Office," *Haaretz*, August 21, 2008.

Wall, Robert, "Fighting Under Fire," *Aviation Week and Space Technology*, Vol. 170, No. 13, March 30, 2009, p. 31.

Warwick, Ned, "Battle-Tested: Young Israelis, Fresh from Front, Recount Clash," *The Philadelphia Inquirer*, July 25, 2006, p. A1.

Wedgwood, C. V., *William the Silent*, London: Cape, 1967.

Wegman, Yehuda, "The Struggle for Situation Awareness in the IDF," *Strategic Assessment*, Vol. 10, No. 4, February 2008.

Weiss, Efrat, "Tyre Raid 'Heroic Operation,'" YNetNews.com, August 5, 2006. As of August 30, 2010:
http://www.ynetnews.com/articles/0,7340,L-3286565,00.html

Weizman, Eyal, *Hollow Land: Israel's Architecture of Occupation*, London: Verso, 2007.

White, Jeffrey, *If War Comes: Israel vs. Hizballah and Its Allies*, Policy Focus No. 106, Washington, D.C.: The Washington Institute for Near East Policy, 2010.

The Winograd Commission—*see* The Commission for the Examination of the Events of the 2006 Campaign in Lebanon.

Wohlgelertner, Elli, "The Mayor's Grand Plan," *The Jerusalem Post*, December 5, 1997, p. 14.

Zabriskie, Phil, "The Former Dove Who's Directing Israel's War," *Time*, July 17, 2006.

Zino, Aviram, "I Wouldn't Have Gone to War, Peres Tells Winograd Commission," YNewsNet.com, March 22, 2007. As of March 15, 2009: http://www.ynetnews.com/articles/0,7340,L-3380005,00.html

Index